Managed Pressure Drilling

Managed Pressure Drilling

Editor

Lalit Verma

scitus
academics

Managed Pressure Drilling
Edited by **Lalit Verma**

Printed in 2017

ISBN: 978-1-68117-411-2

Library of Congress Control Number: 2015941619

© 2016 by
SCITUS Academics LLC,
616, Corporate Way, Suite 2, 4766,
Valley Cottage, NY 10989

www.scitusacademics.com

Contents

Preface

Managed Pressure Drilling Operations is a significant technology worldwide and beginning to make an impact all over the world. Often reservoir and drilling engineers are faced with the decision on how best to construct a well to exploit zones of interest while seeking to avoid drilling problems that contribute to reservoir damage or cause loss of hole. The decision to pursue a MPD operation is based on the intent of applying the most appropriate technology for the candidate and entails either an acceptance of influx to the surface or avoidance of influx into the wellbore. In today's exploration and production environment, drillers must now drill deeper, faster and into increasingly harsher environments where using conventional methods could be counter-productive at best and impossible at worst. Managed Pressure Drilling (MPD) is rapidly gaining popularity as a way to mitigate risks and costs associated with drilling in harsh environments. With this book in hand drilling professionals gain knowledge of the various variations involved in managed pressure drilling operations; understand the safety and operational aspects of a managed pressure drilling project; and be able to make an informed selection of all equipment required to carry out a managed pressure drilling operation.

Editor

Emerging Trend in Natural Resource Utilization for Bioremediation of Oil – Based Drilling Wastes in Nigeria

Iheoma M. Adekunle[1], Augustine O. O. Igbuku[2], Oke Oguns[3], and Philip D. Shekwolo[2]

[1]Environmental Remediation Research Group, Department of Chemical Sciences (Chemistry), Federal University Otuoke, Bayelsa State, Nigeria

[2]Restoration of Ogoniland Project Team, Shell Petroleum Development Company, Port Harcourt, Nigeria

[3]Remediation Team, Shell Petroleum Development Company, Port Harcourt, Nigeria

INTRODUCTION

Background

Nigeria is a country endowed with diverse mineral and natural resources among which is petroleum, a pivot to the national economy and sustainable development. In the past five decades, petroleum exploration and production activities have brought national economic boom but not without some aches. Acts of sabotage such as crude oil theft, pipeline bunkering and artisanal refining added to accidental spills and operational failures all combine to aggravate the oil-related aches. Oil spill into the environment, stemming from either acts of sabotage or operational failures, ultimately lead to environmental pollution with petroleum hydrocarbons [1, 2]. Petroleum mining or drilling is another factor to petroleum hydrocarbons in the environment. Most of the adverse impacts of oil spill/ petroleum hydrocarbons in the environment are experienced in the oil bearing communities, located in the Niger Delta region of the country; prominent among them being the Ogoni land pollution incidence reported by United Nations Environment Programme [1]. Petroleum exploration and production activities are strongly associated with drilling operations for oil mining. Accordingly, the extraction of petroleum resources from the earth is achieved by drilling activities. A developed drilling concept, irrespective of technological advancement, has its technical challenges, process requirements and environmental issues [3]. Drilling fluids, also referred to as drilling muds are used to enhance drilling activities via suspension of cuttings, pressure control, stabilization of exposed rocks, provision of buoyancy, cooling and lubricating.

Types of Drilling Fluids (Muds): There are basically two categories of drilling fluids namely (i) aqueous drilling muds or water based muds (WBMs), which consist of fresh or salt water containing a weighting agent, usually barite ($BaSO_4$), clay or organic polymers and various inorganic salts, inert solids, and organic additives to modify the physical properties of the mud so that it functions optimally and (ii) non-aqueous drilling fluids (NADFs), which comprise all non-water dispersible base fluids such as oil based muds (OBMs) and synthetic based muds (SBMs) [2]. Comparative evaluation of oil based muds and water based muds shows that OBMs offer advantages over WBMs for the reasons that [3]:

- OBMs are more suitable to drill sensitive shells, allowing drilling faster than the WBMs, providing excellent shale stability
- they are more adequate to drill formulations where bottom hole temperatures exceed WBMs tolerance, especially in the presence of contaminants such as water, gases, cement, salt and temperature up to 550F
- OBMs resist formation salt leach out
- they are characterized by thin filter cakes and the friction between the pipe and wellbore is minimized, thus, reducing the risk of differential sticking and are especially suited for highly deviated and horizontal wells
- the drill of low pore pressure formations is easily accomplished, since mud weight can be maintained at a weight less than that of water (as low as 7.5 ppg)
- corrosion of pipe is controlled since oil is the external phase and coats the pipe. The oils are non-conductors and the additives are thermally stable, hence, do not form corrosive products
- bacteria do not thrive long in OBMs
- there is the possibility of using OBMs over and over again and can be stored over long periods of time since bacterial growth is suppressed
- OBM packer fluids are designed to be stable over long periods of time even when exposed to high temperature and provide long-term stable packers since additives are extremely temperature stable. Properly designed, such packer fluids can suspend weighting materials over long periods of times.

In other words, regarding shale stability, penetration rate, high temperatures, drilling salts, lubrication, low pore pressure formations, corrosion control, re-use and packer fluids, OBMs offer advantages over WBMs. It is therefore, obvious that though WBMs are more environmentally benign, they are only satisfactory for less demanding drilling of conventional vertical wells at medium depths, whereas OBMs are more suited for greater depths or in directional or horizontal drillings, which exert greater stress on drilling apparatus. As a result, OBMs are more frequently used in petroleum industries for drilling purposes. The composition of OBMs include: petroleum base fluid, weighting agent and other chemical additives.

Drill Cuttings: During drilling, particles of crushed rocks produced by the grinding action of the drill bit as it penetrates the earth are referred to as drill cuttings (DC). DCs are, therefore, a mixture of rocks and particulates released from geological formulations in the drill holes made for crude oil drilling and are usually coated with the drilling fluid. Consequently, DCs are largely influenced by the chemical composition of drilling muds [2, 4].

The resultant spent OBM and drill cuttings (drilling wastes) consist of hydrocarbons, water, soils, heavy metals and water soluble salts such as chlorides and sulphates [3, 4]. Drilling wastes, which are toxic due to the presence of hydrocarbons, heavy metals and other chemical additives, if not properly treated before disposal, pose serious environmental hazards and risk to public health. Sequel to these, best practices in the management of drilling wastes cannot be over emphasized.

Health and Environmental Effects Associated with Drilling Wastes

Health effects linked to drilling wastes are traceable to the basic components such as the drilling fluid and additives:

Health Effects Associated with Drilling Fluids: These health effects are attributed to the physical and chemical properties of the drilling fluids. In oil based drilling wastes, the base oil stem from petroleum stream such as crude oil, diesel (gasoil) and kerosene, which cause skin irritation. Consequently, the most commonly observed health effect associated with drilling fluids is skin irritation. Other effects include headache, nausea, eye irritation and coughing. Routes of exposure in human are dermal, inhalation, oral and some other miscellaneous routes. On exposure to drilling fluid, petroleum hydrocarbons tend to remove natural fat from the skin, which results in skin drying and cracking. These conditions allow compounds to permeate through the skin leading to irritation and dermatitis. Susceptibility to these health effects varies with individual resistance capacity and conditions of poor personal/environmental hygiene. High aromatic content fluids, especially diesel fuel contain significant levels of carcinogenic polynuclear aromatic hydrocarbons (PAHs). Diesel fuels may also be genotoxic due to high proportions of 3-7 ring PAH [2]. Skin-painting studies in mice showed that, irrespective of the level of PAH, long-term dermal exposure to diesel fuels can cause skin tumours, an effect attributed to chronic skin irritation. In humans, chronic irritation may cause small areas of the skin to thicken, eventually forming rough wart-like growths, which may become malignant. Health effects from chronic exposure to PAHs may include cataracts, kidney damage, liver damage and jaundice. Naphthalene, a specific PAH, can cause the breakdown of red blood cells, if inhaled or ingested in large amounts. Animals exposed to levels of some PAHs over long periods in laboratory studies, developed lung cancer from inhalation and stomach cancer from ingesting PAHs in food [2].

Other hydrocarbon constituents of drilling fluids are the mono-aromatics popularly referred to as BTEX (benzene, toluene, ethylbenzene and xylene). BTEX compounds are very volatile, hence, will readily evaporate in warm/hot climates of tropical regions, resulting in higher concentrations in the vapor phase. As a result, there is the possibility of exposure to human via inhalation. Exposure to high concentrations of these hydrocarbons via inhalation may result in hydrocarbon induced neurotoxicity, a non-specific effect resulting in headache, nausea, dizziness, fatigue, lack of coordination, problems with attention and memory, gait disturbances and narcosis [2].

Health Effects Associated with Additives: In addition to the irritancy of the drilling fluid hydrocarbon constituents, several drilling fluid additives may also have irritant, corrosive or sensitizing properties. Various additives include emulsion stabilizers, pH adjusters, wetting agents, viscosifiers and fluid-loss reducing agents. For instance, calcium chloride ($CaCl_2$) has irritant properties and emulsifiers (such as polyamine) have been associated with sensitizing properties [3]. Specific chemical additives vary with locations.

Environmental Effects Associated with Drilling Wastes

Apart from health effects, environmental hazards associated with drilling wastes include land, water and air pollution [5]:

- *Land pollution*: Farming is the major land use system in Nigeria, especially in the Niger Delta region [1]. The most significant in this aspect of environmental pollution in Nigeria is thus farmland pollution. Consequences include alteration in soil physical, biological and chemical properties, loss of soil fertility, stunted plant growth and reduced crop productivity. These lead to reduced food security and compromised food safety.

- *Aquatic pollution*: Large percentage of the oil spill gets spread over the surface of the aquatic system resulting in anaerobic environment in the water, below the surface. This leads to death of the natural flora and fauna where oxygen is the key element for their respiration; adversely affecting fishing profession [1]

- *Air pollution*: volatile organics such as benzene, toluene, ethylbenzene and xylene could have elevated concentrations in the air, leading to atmospheric pollution and consequent adverse environmental and health impacts.

Oil well drilling processes generate large volumes of drill cuttings and spent mud in the country. Drilling wastes, therefore, add to hazardous petroleum waste materials released in the environments of the Niger Delta region of the country [1, 6] and the management of drilling wastes is quite tasking. An environmentally friendly technique for the management of drilling wastes is necessary in all offshore and onshore operations; from seismic surveys, drilling operations, field development and production to decommissioning. The physical and chemical properties of the drilling wastes influence their hazardous characteristics and environmental impact abilities, which in turn depend primarily on: (i) nature of impacted material, (ii) concentration of pollutant /amount of waste material after release (iii) recipient biotic community and (iv) exposure duration. Exposure that causes an immediate effect is called acute exposure while long-term exposure is called chronic exposure. Either acute or chronic exposure has negative impacts.

Contemporary Treatment of Drilling Waste Materials

Worldwide, contemporary drilling waste management options include re-use, offshore discharge, re-injection and onshore treatment and/or disposal [7]. Each treatment and or disposal option has its pros and cons as highlighted in the options (thermal technologies and bioremediation techniques) discussed.

Thermal Treatment

As the name suggests, thermal technologies involve the use of high temperatures to reclaim hydrocarbon contaminated materials [8]. Thermal treatment is mostly used in treating organic compounds. Additional treatment may be necessary for metals and salts depending on the final fate of the wastes. Thermal treatment technologies are designed for a fixed land based installation; however, a few mobile units exist. Two commonly practiced thermal treatment technologies are thermal desorption and incineration methods.

Thermal Desorption Method

Thermal desorption is an environmental remediation process that uses heat to increase the volatility of contaminants by the use of a series of equipment (desorber and oxidizer) such that the hydrocarbons and water are separated or removed from the solid matrix. It is normally carried out between the temperature range of 250-650°C. At these temperatures both the lighter and heavier hydrocarbons are removed and collected or thermally oxidized by further heating to a temperature of over 850°C. The resulting solid residue has essentially no residual hydrocarbons (having been oxidized), but does concentrate salts and heavy metals. Depending upon the success of process used, recovered hydrocarbons can be used as fuel or re-used as base fluid in the drilling fluid system and the resulting solid can be disposed of in a landfill or may be used in construction (of roads and bricks). Economical, operational and environmental implications of thermal desorption include:

- Effective removal and recovery of hydrocarbons from solids
- Possibility of recovering base fluid and end - product could be used for brick making
- Low potential for future liability
- Requires short time
- High cost of handling environmental issues
- Large volume of wastes is required to justify the cost of operation
- Requires tightly controlled process parameters
- High operating temperatures can lead to safety risks
- Requires several operators
- Heavy metals and salts are concentrated in residual solids
- Process water contains some emulsified oil
- Residue ash requires further treatment before disposal
- End product is sterile and can no longer support plant Life.

Incineration Method

Incineration involves (i) heating oil based mud and drill cuttings to a higher temperature range (1200-1500°C) in direct contact with combustion gases and (ii) oxidizing the hydrocarbons [8]. Solid/ash and vapor phases are generated. The gases produced from this operation may be passed through an oxidizer, wet scrubber, and bag house before being vented to the atmosphere. Stabilization of residual materials may be required prior to disposal to prevent constituents from leaching into the environment. Incineration of drilling wastes occurs in rotary kilns, which incinerate any waste regardless of size and composition. Incineration systems are designed to destroy only organic components of waste; however, most drilling wastes are non-exclusive in their content and therefore will contain both combustible organics and non-combustible inorganic materials. By destroying the organic fraction and converting it to carbon (IV) oxide and water vapor, incineration reduces waste volume. Inorganic components of wastes fed to an incinerator cannot be destroyed, only oxidized. The major inorganic materials are chemically classified as metals. Generally, these metals will exit the combustion process as oxides of the metals that enter. Economical, operational and environmental implications of incineration are as listed:

- Low potential for future liability
- High cost per volume
- Heat produced could be used for energy generation
- High energy cost
- Requires air pollution control equipment because of safety concerns
- At high temperatures, salts can form acid components
- Air emissions pose environmental concerns.

In line with best practices, for thermal technologies, there is need for proper placement of end product. Demonstration of sufficient compliance with current regulations and adequate safety measures to cater for the potential risks of exposure to high temperatures.

Bioremediation Technique

Bioremediation technique relies on the ability of microorganisms (mostly combination of bacteria) to feed on the hydrocarbons (HCs) as substrate, converting them into carbon dioxide, water and harmless clean solids; and the ability of some of the HCs to biodegrade over time. But in most cases, the native microorganisms are often overwhelmed by the extent of the hydrocarbon contamination and thus would require external nutrients to boost (bio-stimulation) their activity and ability to take up the HCs at a faster rate. In other cases, the native microorganisms may be needing help from their kind or other species of micro-organism which are grown or inoculated (bio-augmentation) in the laboratory and then introduced in the habitat of the native micro-organisms. Bioremediation could be carried out at the site of contamination (in-situ bioremediation technique) or off the site of contamination (ex-situ bioremediation technique). Bioremediation technologies include land farming, use of bioreactors, biopiles and compost-based technologies. Economical, operational and environmental implications of conventional bioremediation technique [9, 10, 11, 12, 13, 14] include:

- Relatively inexpensive
- Requires simple equipments and eliminates transportation cost as drill wastes could be treated on site
- Less capital but may be labour-intensive.
- Low maintenance cost; being a simple technology process that requires few machines, there are few delays due to equipment down-time
- Process is fairly flexible and can be used for most drill wastes including OBM, NADFs, previously extracted materials and newly drilled cuttings
- Proven technology
- Requires a considerable period of time to complete a process
- Appropriate bacteria and nutrient selection could be a daunting task

- In cases where bacteria are inoculated and brought on site, adaptability to their new environment may hamper their performance
- Minimal operation hazards
- Environmentally friendly: once the contaminants have been degraded, the microbial population reduces considerably as they have used up their food source
- Less impact on the environment as residue from process (TPH < 1%) may require no further treatment and could be used for agricultural purposes.

Recommended best practices for bioremediation technology include ensuring (i) proper initial physical, biological and chemical characterizations to determine extent of organic and inorganic contamination, (ii) required skill and persistence for the selection of several combinations of bacteria and nutrients that can provide the desired result (iii) proper periodic tillage to provide for proper aeration that facilitates degradation of the HCs and (iv) an accurate and appropriate TPH level check in between treatment process in order to monitor progress of the remediation process. Choice of waste management options typically considers local regulations, environmental assessment, cost/benefit analysis and the composition of the drilling wastes. The Department of Petroleum Resources [15] via the Environmental Guidelines and Standards for the Petroleum Industry in Nigeria (EGSPIN) stipulated guidelines on drill cuttings discharge for inland / near-shore and offshore deep water in order to minimize the adverse impact on the surrounding environment. These requirements call for an appropriate drill cuttings treatment prior to disposal in order to meet the stipulated conditions.

Review of Emerging Trend in the Treatment of Drilling Waste Materials in Nigeria

There are scientific evidences showing that drilling wastes generated in the country contain toxicants that are of environmental concerns. For instance, the reports of [16] on the determination of selected physical and chemical parameters including metals concentrations in a certain drill cutting dump site in the country. Results from their study showed that oil and grease on the surface and 20 feet around the waste dump area were above the specified limit [15]. There was also lack of plant growth noticed in the study, attributed to depletion of nitrogen, phosphorus and potassium values below threshold levels for plant growth. The reports of [4] on hydrocarbon and some metal contents of drilling muds and cuttings generated during the drilling of Igbokoda onshore oil wells gave total petroleum hydrocarbon (TPH), aliphatic hydrocarbon (AH) and polycyclic aromatic hydrocarbon (PAH) as generally exceeding stipulated limits by both national and international agencies. The studies of [17] on the compositional distribution and sources of polynuclear aromatic hydrocarbons (PAHs) in Nigerian oil-based drill-cuttings, showed that the total initial PAHs concentration of the drill cuttings was 223.52 mg/kg while the initial individual PAHs concentrations ranged from 1.67 to 70.7 mg/kg, dry weight, with a 90% predominance of the combustion-specific 3-ring PAHs.

The commonly employed remediation techniques for drilling wastes in Nigeria appear to be thermal technologies. However, due to economical, operational and environmental implications of these thermal technologies; search for more acceptable techniques commenced. There is scarcity of literature on the use of natural resource materials for the remediation of drilling wastes in Nigeria. The few literature resources showed that a large percentage is still at the bench-scale platform. For instance, [18] isolated*Staphylococcus sp*. from oil-contaminated soil that was treated with 1% drilling fluid base oil (HDF-2000). Their study revealed that *Staphylococcus sp.*, is a strong primary utilizer of the base oil and has potential for application in bioremediation processes involving oil-based drilling fluids. On the other hand, the effectiveness of 2 bacterial isolates (*Bacillus subtilis* and *Pseudomonas aeruginosa*) in the restoration of oil-field drill-cuttings contaminated with polynuclear aromatic hydrocarbons was studied by [19]. In that study, a mixture of 4 kg of drill cuttings and 0.67 kg of top-soil were fed into triplicate plastic reactors labeled A1 to A3, B1 to B3, C1 to C3 and O1 to

O3. These were left quiescent for 7 days under ambient conditions, followed by the addition of 20 mL working solution of pure cultures of*Bacillus* sp and *Pseudomonas* sp (each of cell density 7.6 x 10^{11} cfu/mL) to reactors A1 - A3 and B1 - B3 respectively. Another 20 mL working solution containing both cultures at cell density 1.5 x 10^{12}cfu/mL was added to reactors C1 - C3. The working solution was added to each reactor (excluding the controls, O1 - O3) every 2 weeks. Mixing and watering of the set-ups were carried out at 3 days interval under ambient temperature of 30°C for a period of 6 weeks. Results showed that the predominant 3-ring PAHs, which made up 90% w/w of the total PAHs concentration of 223.52 mg/kg, were degraded below detection and the 4-ring PAHs were reduced from 4 to 0.6% by *Pseudomonas* while *Bacillus*reduced 3 and 4-ring PAHs respectively to 0.2 and 0.8%. Their works revealed that Pseudomonas degraded 3 and 4-ring PAHs relatively better than *Bacillus*. Both strains of bacteria degraded 5 and 6-ring PAHs below detection limits. Furthermore within the 3-ring PAHs, each of the strains of bacteria reduced phenanthrene to approximately 0.2%, whereas both degraded homologues acenaphthylene, acenaphthene and fluorene as well as anthracene below detection limits. For 4-ring PAHs,*Pseudomonas* degraded fluoranthene and benzo[a]anthracene. *Bacillus* also degraded benzo[a]anthracene below detection limits. *Pseudomonas* was able to reduce pyrene and chrysene to 0.3 and 0.2% respectively; whereas *Bacillus* reduced fluoranthene, pyrene and chrysene to 0.1, 0.01 and 0.4% respectively. However, treatment with the mixed culture resulted in limited degradation of 5-ring PAHs particularly in the fourth week, which was attributed to the phenomena of co-metabolism and inhibition.

The works of [20] compared the potentials of bio-augmentation and conventional composting as bioremediation technologies for the removal of PAHs from oil-field drill-cuttings. From a mud-pit, close to a just-completed crude-oil well in the Niger Delta region of Nigeria, 4000 g of drill cuttings was obtained and homogenized with 667 g of top-soil (to serve as microbes carrier) in three separate reactors (A, B and C). The bio-augmentation of indigenous bacteria in the mix was done by adding to reactors A and B a 20-mL working solution (containing 7.6×10^{11} cfu/mL) of pure culture of *Bacillus*and *Pseudomonas,* respectively, while a 20-mL working solution (containing 1.5×10^{12} cfu/mL) of the mixed culture of *Bacillus* and *Pseudomonas* was added to reactor C. The bio-preparation was added to each reactor (excluding the control) every two weeks for six weeks. The composting experiment was conducted in a 10-litre reactor in which 4000 g of drill cuttings, 920 g of topsoil and 154 g of farmyard manure and poultry droppings were homogenized. Mixing and watering of the set-ups were carried out at 3 days interval under ambient temperature over a period of six weeks. Results showed that initial individual PAHs concentrations in the drill cuttings ranged from 1.67 to 70.7 mg/kg dry weight, with a predominance of combustion-specific 3-ring PAHs (representing 90% of a total initial PAHs. After the bioremediation exercise that lasted for 42 days, total PAHs in the drill cuttings were reduced from 223.52 to 4.25 mg/kg, representing a 98.1% reduction. Away from the use of microbial strains in the treatment of drilling wastes, a bench-scale investigation was carried out by [21] to demonstrate the efficacy of technique referred to as 'Dispersion by Chemical Reaction (DCR) technology".This particular method involved the use of hydrophobized calcium oxide (CaO) to form a dry, soil-like material that could be useful in construction works.

On the other hand, after the study on the response of four phytoplankton species in some sections of Nigeria coastal waters to crude oil in controlled ecosystem [22], that revealed the adverse impacts; a multidisciplinary environmental remediation research group (ERRG) was inaugurated with the mandate to embark on innovative, cutting-edge research and development (R & D) initiative, aimed at the development of an indigenous technology for an eco-friendly technique in the treatment of soils, sediments, sludge and drilling wastes polluted by petroleum hydrocarbons, using natural products of Nigeria origin. The goal of ERRG is to translate the technology from bench-scale to field scale and come out with on- the - shelf products that will find use for both onshore and offshore remediation works. The first phase of the R

& D initiative was the exploration of the remediation potential of conventional composting technology based on the results from the works of [23]. A good start was the production of a scientifically formulated and classified compost bulk [24] that are potentially viable for environmental remediation projects [25] and able to biodegrade petroleum hydrocarbons embedded in soil and related matrices [26]. The next phase was to assess public acceptance of the principles of this technology, which culminated to the reports of [27] on population perception impact on value-added solid waste disposal in developing countries, a case study of Port Harcourt City. The feedstock utilized in product formulations in this emerging, indigenous and innovative technology is 100% biodegradable and very abundant in the Nigerian environment. Consequently, the technology has been categorized by stakeholders [27] as:

- eco-friendly environmental remediation technique
- waste to wealth initiative
- waste to resource initiative
- value-added waste management option
- a contribution to the promotion of local material development that has the potential for:
  - wealth creation
  - job creation
  - poverty alleviation
  - sound environmental management of hydrocarbon polluted wastes from the petroleum industries.

ERRG observed that either conventional composting technology or bioremediation via utilization of pure microbial isolates/strains has limitations in terms of serving the practical needs of the petroleum industry in Nigeria with regards to meeting (i) regulatory remediation targets at close – out of project and (ii) project delivery time. Subsequently, through series of bench-scale and screen house remediation investigations, products were formulated to enhance the speed of bioremediation process using nano-scale green catalysts, a technique that matured into Compost - based Nanotechnology in Bioremediation (CNB-Tech). The research group then subjected the CNB-Tech products to different scientific evaluations in order to ascertain (i) efficiency on biodegradation of petroleum hydrocarbons in oily wastes such as crude oil impacted soils, sludge and drilling wastes (drill cuttings and oil-based mud) and (ii) environmental impacts with emphasis on soil quality. Published works on assessment and prognosis of products' impact on soil quality include:

- Assessing the effect of bioremediation agent from local resource materials in Nigeria on soil pH [28]
- Impact of bioremediation formulation from Nigeria local resource materials on moisture contents for soils contaminated with petroleum [29]
- Assessing and forecasting the impact of bioremediation product derived from Nigeria local raw materials on electrical conductivity of soils contaminated with petroleum products [30]
- Soil temperature dynamics during bioremediation of petroleum products using remediation agent from Nigerian local resource materials [31].

Other works on CNB-Tech products' evaluations including (i) effect on soil heavy metal dynamics and (ii) impact on soil microbial species population and diversity are being considered elsewhere for publication. Having recorded a huge success during the laboratory scale investigations where maximum of 4000g of sample bulk and freshly hydrocarbon contaminated soils (similar to the quantities used by other investigators) [19, 20] were treated, it became necessary to assess the efficiency of CNB-Tech products on waste materials with complex nature and higher degree of hydrocarbon pollution. This aspiration was realized in collaboration with the Remediation Department of Shell Petroleum Development Company (SPDC), Port Harcourt, Nigeria through the University Liaison Team of SPDC. Sequel to this, pilot-scale projects were commissioned to evaluate the efficiency of CNB-Tech products on the degradation of hydrocarbon compounds in the following petroleum impacted materials:

- Hydrocarbon polluted clay soils from Ejama-Ebubu legacy site of SPDC
- Hydrocarbon polluted carbonized soil from Ejama-Ebubu legacy site of SPDC
- Hydrocarbon polluted sludge from Ejama-Ebubu legacy site of SPDC
- Oil-based mud and drill cuttings generated from SPDC operations.

Ejama Ebubu is one of SPDC's legacy sites of up to 42 year long pollution as at the time of study in 2011 [1]. In this chapter, the efficacy of CNB-Tech products in the biodegradation of petroleum hydrocarbons in oil-based drilling wastes (OBM-DC) is presented.

Research Justification

The treatment of drilling wastes, especially OBM-DC in an environmentally sound manner is a challenging task due to the complex nature of the wastes. The most popular technique adopted for the treatment of OBM-DC, thermal desorption [15] has its accompanying environmental concerns. For instance, thermal treatment technologies are associated with prohibitive capital and operational cost implications, threatening environmental consequences in addition to high occupational hazards and generation of secondary waste stream that has to be treated at extra high cost before final disposal. Consequently, there is need for a pragmatic shift to seek alternative techniques that will address the need of the oil and gas sector in the management of drilling wastes in terms of remediation target delivery time and compliance to regulatory standards in Nigeria. Regulatory standards for close-out of remediation projects vary from one country to another and success factors of a given technology are dependent on indices such as:

- climatic conditions

- geographical characteristics of the location
- nature and complexity of contamination
- expected utility of the end-products of the remediation exercise

It then becomes evident that a successful remediation technology in one part of the globe may not necessarily be efficient in another region, pointing to the need to look inward for a more practical approach to solving the environmental challenges posed by petroleum hydrocarbon polluted waste streams in Nigeria [1]. Having run laboratory, bench- scale and screen-house remediation works using CNB-Tech products on fresh hydrocarbon contaminated soils, it became necessary to conduct pilot scale remediation works on more challenging waste streams such as weathered petroleum impacted soils, sludge, sediment, oil- based drilling mud and drill cuttings, hence this project.

Research Objectives

The current study comprised three major objectives:
- to conduct a review on the emerging trends in the treatment and related studies for drilling wastes in Nigeria,
- to assess the efficiency of an indigenous and innovative application of compost - based nanotechnology in bioremediation (CNB-Tech) in biodegradation of hydrocarbons found in oil-based mud and drill cuttings; generated by a petroleum industry in Nigeria
- to investigate the beneficial utility of the remediation end-product for agricultural purpose (crop production), which is a major land use system in Nigeria.

RESEARCH METHODOLOGY

The research methodologies employed in this study were:
- Literature review to provide an insight to the current and

emerging trend in the treatment of drilling waste materials in the country and

- Practical, ex-situ, pilot scale execution of biodegradation of hydrocarbon compounds in oil-based mud and drill cuttings generated by an oil company in Nigeria using an indigenous and innovative biotechnological (CNB-Tech) approach anchored on the use of natural resource materials of Nigeria origin.

Pilot-scale Remediation of Oil-based Mud and Cuttings Using CNB-Tech Method

This study was carried out during the 2010/ 2011 Sabbatical Programme of the University Liaison Team of Shell Petroleum Development Company (SPDC); in conjunction with the Remediation Department of SPDC, Port-Harcourt, Nigeria. The indigenous remediation products (CNB-Tech products) prepared from cellulosic natural resource materials and biogenic nanopolymers of Nigeria origin used for this pilot remediation study, were denoted as (i) Ecorem, (ii) Bioprimer and (iii) Biozator. The last two products are solids that are transformed to the aqueous form before use while the first product is used in the solid form.

Project Site Description

The present pilot-scale project, for the purposes of adequate monitoring and efficient execution, was carried out in the Industrial Area of Shell Petroleum Development Company, Port Harcourt, Rivers State; known as "Shell IA". The earmarked project area was a relatively isolated open green field within Shell IA and according to design, a temporary sheltered facility constructed to suit the project design was erected at the site and all necessary health and safety issues were taken into consideration. The sheltered project facility comprised of three major units:

- Remediation execution section: where actual remediation took place

- Phyto-analytical section: where effects on plant life were investigated
- Mini- chemical laboratory: where necessary onsite chemical evaluations were conducted.

Pilot Scale Remediation Procedure

The batch of oil-based mud and drill cuttings (OBM-DC) used in this study was generated from SPDC's operations and supplied by one of the company's certified vendors. During the conveyance procedure for OBM-DC, chain of custody document and waste stream tracking manifest was observed. Basic highlights for CNB-Tech application mode are outlined in Figure 1. Pretreatment involved recovery of free phase base fluid and stabilization involved modification of viscosity parameter.

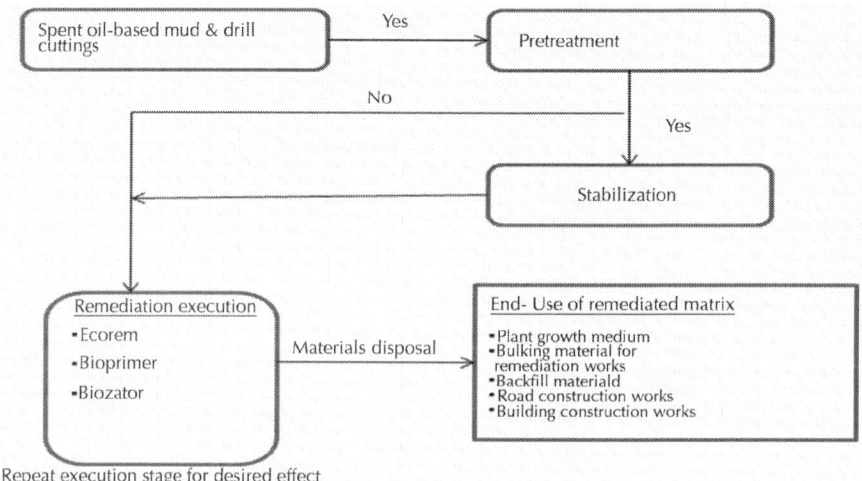

Figure 1: Application model of CNB-Tech remediation method.

The biocell utilized for the remediation execution was designed by the research group, locally fabricated and lined with appropriate PVC materials. The procedures involved in the pilot remediation exercise are described as follows: A biocell of total dimension 15 m³ was sub-divided to smaller units of 3 m x 1 m x 1 m to allow for five times replication. Ecorem (a CNB-Tech product) was placed in the cells prior to loading of oil-based drilling mud and cutting (OBM-DC) that have been previously conditioned using intervention CNB-Tech products. As the initial microbial population in OBM-DC was less than 2.0×10^3 cfu/mL, Ecorem was introduced at 10% by weight of waste materials. Using mechanical means, OBM-DC and Ecorem were homogenized and allowed to incubate for about 12 to 24 hours in order to trigger and stimulate natural microbial activities. CNB-Tech products (Bioprimer and Biozator) were then applied to saturate the contents in the biocells, which was followed by homogenization using mechanical devices. A CNB-Tech product was added to the leachate (process fluid) to immobilize inorganic constituents (especially metals) before recycling the leachate into the treatment network in such a manner that no leachate was produced as a by-product for discharge into the environment. OBM-DC that received no treatment served as control. Both controls and test units were subjected to the same environmental conditions.

System maintenance and monitoring: During remediation, the system was monitored for relevant environmental factors such as moisture content (I), pH (II), nitrogen content (III) and temperature (IV) using standard procedures of gravimetry for I, probe method via a calibrated pH meter for II, Kjedahl method for III and calibrated mercury in glass thermometer for IV. These environmental factors were maintained at the required range. Remediation lasted for 33 days: 6 days for actual treatment and 27 days for material fallow and recovery periods during which the treated materials were conditioned with a CNB-Tech product (Ecorem) for use as plant growth medium. In order to validate the efficacy of this technology, representative composites were sent to an International Laboratory (RespirTeK Consulting Laboratory and affiliate Laboratories based in the United States of America) for physical, chemical and microbial assessments. RespirTek is ISO/EC accredited and certified. Three other laboratories that are based in Nigeria (certified by national regulatory bodies) were also involved in sample collection and analyses. Laboratories that participated in this study were:

- Technology Partners International Nigeria Limited, Port Harcourt - Nigeria
- Laser Engineering and Resources Consultants Limited, Port Harcourt- Nigeria
- Fugro Nigeria Limited, Port Harcourt, Nigeria
- RespirTek Consulting Laboratory - United States of America

Sample Collection

At the end of the pilot remediation project using CNB-Tech products, treated materials were moved from the biocells and spread out on PVC impermeable membranes (each of dimension 650 cm for length and 248 cm for width), homogenized using mechanical means and air-dried with occasional homogenization of samples. The dry samples were returned into the biocells where further homogenization procedure was carried out. Sampling containers were sent by RespirTEK Consulting Laboratory, USA for their own use.

General Sample Collection: Using mechanical means, treated and dried samples in the cells were thoroughly homogenized for one week. In order to collect sample from a particular replicate, each replicate was subdivided into 4 equal parts; representative fractions were collected from the different parts and recombined to give a composite sample of 1kg.

BTEX Sampling: Standard sampling kit for BTEX, sent by RespirTEK Consulting Laboratory, was utilized for the purpose. In this procedure, homogenized samples were collected from the cells using "Terra Core" sampling device. Using a 40 mL glass VOA vial containing appropriate preservatives and with the plunger seated in the handle, the Terra Core was pushed into freshly homogenized sample until the sample chamber was filled to the capacity of 5g. All sample particulates (debris) were removed from the outside of the Terra Core sampler and the sample plug was pushed into the mouth of the sampler. Excess soil that extended beyond the mouth of the sampler was removed. The plunger was then seated in the handle and rotated until it aligned with the slots in the body. The mouth of the sampler was placed into the 40 mL VOA vial containing the preservatives and sample extruded by pushing the plunger down. The lid was quickly placed back on the 40 mL VOA vial. It was ensured that when capping the 40 mL VOA vial, sample debris was removed from the top of the vial.

All samples were appropriately labeled and recorded in the chain of custody form before shipping to the USA laboratory by courier. Two Laboratories in Nigeria also collected samples for analyses, following standard procedures. The third laboratory in Nigeria was only involved in the analysis of materials using infrared and UV-absorption spectroscopic methods.

Physicochemical Analysis and Microbial Assessment

Statement from quality control and quality assurance unit (QA/QC) of RespirTek Laboratory, USA showed that all analyses were conducted following procedures set forth by the ISO/IEC 17025:2005 accreditation program standards for which the laboratory holds certification. Quality assurance systems and quality control criteria were strictly followed. The following parameters were determined:

- Total petroleum hydrocarbons (TPH)
- Monoaromatic hydrocarbons: benzene, toluene, ethylbenzene and xylene (BTEX). For xylene, ortho -, meta - and para-derivatives were assessed
- PAHs: a total of 17 PAH compounds: (i) naphthalene, (ii) acenaphthylene, (iii) acenaphthene, (iv) fluorene, (v) phenanthrene, (vi) anthracene, (vii) fluoranthene, (viii) pyrene, (ix) benzo (a) pyrene, (x) chrysene, (xi) benzo (b) fluoranthene, (xii) benzo (k)fluoranthene, (xiii) benzo (a) pyrene, (xiv) dibenz(a,b) anthracene, (xv) benzo (ghi)perylene, (xvi) 2-methylnaphthalene and (xvii) indeno (1,2,3-cd) pyrene
- Metals: barium (Ba), calcium (Ca), copper (Cu), lead (Pb), mercury (Hg), Nickel (Ni), Sodium (Na), Potassium (K),

cadmium (Cd), zinc (Zn) and arsenic (As), a metalloid
- Miscellaneous parameters: pH, salinity, nitrogen, phosphorus, total organic carbon and electrical conductivity.
- Microbial activity: assessment of 48 hr and 96 hr microbial activities of both remediation end-product and contaminated material (control) was conducted by the USA based laboratory. Total hydrocarbon utilizing bacteria as well as total microbial count were assessed by the Nigerian based laboratories.

Hydrocarbon compounds were analyzed using Gas chromatographic method, microbial assessment was carried out using heterotrophic plate count method and metals were determined using atomic absorption spectroscopic technique. All the other parameters were carried out using standard procedures such as described in [24, 25, 32]. The CNB-Tech products (Bioprimer and (Biozator) were characterized using infrared and UV-visible spectroscopic methods. The basic characteristics of Ecorem have already been reported in [24, 25] but was slightly enhanced, in this study, for case specificity.

Assessment of Seed Germination Potential of Treated Samples

The remediated materials used in this evaluation were not mixed with external soil and no external fertilizer material was added to the remediated soil. Seed germination potential (SGP) of treated samples were assessed and only viable maize seedlings were used for this purpose. In a remediated material matrix (4kg material contained in an experimental plastic pot), 6 seedlings of maize were sown. This was replicated three times. All together, 18 (6 x 3) seedlings were used to evaluate this effect. Similar set- ups were also established for the untreated oil – based mud and cuttings, which served as control systems. This gave a total of 18 (6 x 3) seeds tested for germination potential for the test systems and 18 seedlings for the control media. This phase of the evaluation lasted for 7 days.

ASSESSMENT OF PROCESS FLUID (LEACHATE) EFFECT ON PLANT GROWTH

Adequate leachate (process fluid) management strategy was put in place as leachate generated during remediation was recycled into the remediation process. However, this evaluation was to ensure or to prove that in the event of any leachate seepage there would be reduced environmental risk. This phytotoxicity assessment was carried out using a cereal (corn: Zea mays L.,) as an indicator crop and indices of toxicity were (i) root length and (ii) plant height. Experimental systems constituted of the following set-ups, where FS is dilution factor and SF stands for farm soil:

- Farm soil + tap water (Code: FS + water). This served as control system for (ii) and (iii)
- Farm soil + stock leachate (Code: FS + LDF-0). This served as control system for (iii)
- Farm soil + diluted leachate series:
  a. Farm soil + leachate DF-1 (Code: FS + LDF-1)
  b. Farm soil + leachate DF-2 (Code: FS + LDF-2)
  c. Farm soil + leachate DF-3 (Code: FS + LDF-3)
  d. Farm soil + leachate DF-4 (Code: FS + LDF-4)

For this assessment, bulk farm soil sample, obtained from a village (K-dere, part of Ogoniland) in Rivers State, was used. Soil was sieved through a mesh and transferred at 1.5 kg per pot and designated pots were treated to 70% approximate field capacity (determined against gravity) using equal volume of appropriate fluid (water, stock leachate or diluted leachate). The systems were allowed to stabilize for 2 weeks after which viable maize seedlings were sown at 3 per pot. As the plants grew, the soil systems were treated with equal volumes of the appropriate fluid to maintain appropriate moisture level, as required by plant. Experiment lasted for 2 weeks, at the end of which the heights were recorded and plants harvested. Caution was exercised to ensure that roots were not destroyed during harvest. Root lengths were then recorded and mean values per pot calculated for each parameter.

Evaluation of Beneficial Utilization of End-product

Similar to the case in Section 2.4, in this evaluation, the remediated matrix was not mixed with any type of soil, neither was any external fertilizer administered. At close - out of the pilot-scale remediation project, the remediated materials were air dried, primed with one of CNB-Tech products (Ecorem) at a specified loading scheme and then utilized as a growth media. Primed end-products were transferred at 4 kg per pot of 4 liter capacity. Three indicator crops used for this project were:

- Corn (*Zea mays L.,*)
- Green leafy vegetable (Fluted pumpkin: *Telfairia Occidentalis*)
- Cassava (*Manihot esculenta Crantz*)

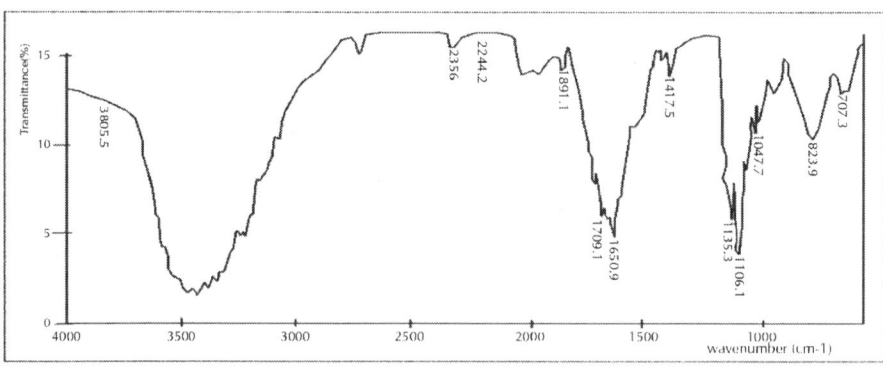

Figure 2: Infrared spectrum of Bioprimer, a CNB-Tech remediation product.

The crops were used because they are commonly grown and consumed in the Niger Delta region of the country. Due to time constraint, duration of investigation varied for the crops, the longest being up to 130 days for green leafy vegetable (Fluted pumpkin: *Telfairia Occidentalis*) while corn (*Zea mays L.,*)and cassava (*Manihot esculenta Crantz*) were grown for 2 and 3 weeks respectively. Untreated OBM-DC served as a control and farm soil served as a second control.

Statistical Analysis

Data generated in this study were subjected to statistical evaluations using SPSS software for Windows, version 17.0. Descriptive statistics were applied to evaluate mean and standard deviation. Paired sample T-Test and One-way analysis of variance (ANOVA) were applied to identify significant variations among treatments as appropriate. Pearson correlation was used to ascertain significant relationships.

RESULTS

Typical Infrared Spectra of Two CNB- Tech Remediation Products

The infrared absorption spectra of two CNB-Tech products (Bioprimer and Biozator) utilized in this pilot scale study are presented in Figures 2 and 3. Both spectra showed absorption peaks in the region of 4000 to 600 cm^{-1}.

Major information from the infrared spectra were: strong, broad absorption band of oxygen-hydrogen (O-H) of an alcohol (aryl/aliphatic) and N-H absorption bonds around 3500 - 3300 cm^{-1}; carbon-oxygen double bond (C=O) absorption band found around 1750 – 1500cm^{-1} This could be carbonyls of ester (RCOOR), aldehyde (RCHO), ketone (RCOR) and acid (RCOOH). C-N bond of nitrogenous matter falls in the end of the range; C-O bond around 1200 – 1000 cm^{-1} and of carbon-hydrogen (C-H) bond for aromatic moieties found below 1000cm^{-1} [33].

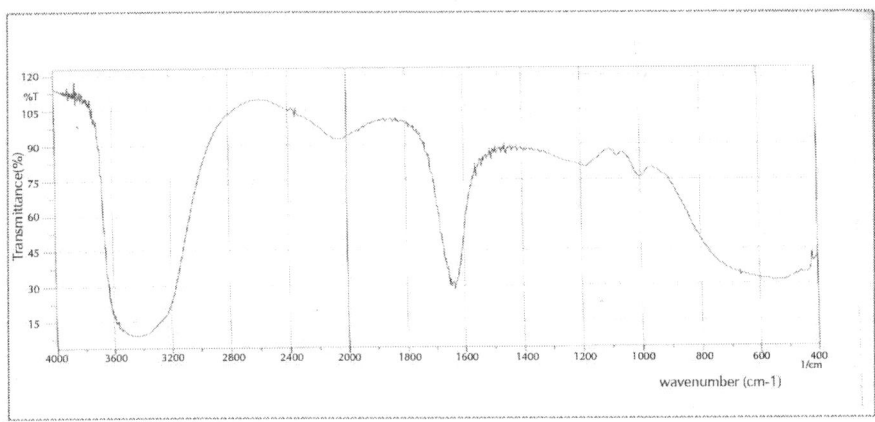

Figure 3: Infrared spectrum of Biozator, a CNB-Tech remediation product.

Initial Characteristics of the Drilling Wastes

The results presented in this paper were largely those obtained from the International laboratory. Table 1 contains the initial characteristics of the drilling wastes (oil-based mud and cuttings).

Table 1: Initial characteristics of the oil -based drilling mud and cuttings used in this pilot scale study

S/N	Parameter	Concentration
Inorganics		
1.	Arsenic (mg/kg)	6.69
*2.	Cadmium	Not determined
3.	Barium(mg/kg)	765
4.	Calcium(mg/kg)	87300
5.	Copper(mg/kg)	35.90
6.	Lead(mg/kg)	161

7.	Mercury(mg/kg)	0.036
8.	Nickel(mg/kg)	12.3
9.	Sodium(mg/kg)	493
10.	Potassium(mg/kg)	1930
11.	Zinc(mg/kg)	144
12.	TKN (%)	0.0357
13.	Phosphorus (%)	0.0291
*14.	pH	10.2
*15.	Electrical conductivity (mSm^{-1})	Not determined
16	Total organic carbon (%)	Not determined
17..	Salinity (mg/kg)	4300
BTEX compounds		
1.	Benzene	0.0198
2.	Ethylbenzene	0.827
3.	m- and p-xylene	0.532
4.	o-xylene	0.924
5.	toluene	1.910
PAH Compounds		
1.	Naphthalene(mg/kg)	1.94
2.	Acenaphthylene(mg/kg)	BDL
3.	Acenaphthene(mg/kg)	BDL
4.	Fluorene(mg/kg)	2.54
5.	Phenanthrene(mg/kg)	0.78
6.	Anthracene(mg/kg)	BDL
7.	Fluoranthene(mg/kg)	BDL
8.	Pyrene(mg/kg)	BDL
9.	Benzo (a) anthracene(mg/kg)	BDL
10.	Chrysene(mg/kg)	BDL
11.	Benzo(b)fluoranthene(mg/kg)	BDL
12.	Benzo (k)fluoranthene(mg/kg)	BDL
13.	Benzo(a)pyrene(mg/kg)	BDL
14.	Dibenz(a,h)anthracene(mg/kg)	BDL
15.	Benzo(g,h)perylene(mg/kg)	BDL
16.	2-methylnapthalene(mg/kg)	5.39
17.	Indeno(1,23-cd)pyrene(mg/kg)	BDL
	Total PAH(mg/kg)	10.65

Total petroleum hydrocarbon		
1.	TPH (mg/kg)	79 200

[i] - *Parameters not determined by the USA laboratory but quantified by Nigerian based laboratories

Results indicated the presence of inorganic constituents and organics (hydrocarbons compounds). Regarding inorganics, soft metal contents increased in the order: Na (493 mg/kg) < K (1930 mg/Kg) < Ca (87, 300 mg/kg). The elemental ratios were 177 for Ca/Na, 45 for Ca/K and 4 for K/Na. Heavy metal concentrations increased in the order: Hg < As < Ni < Zn < Cu < Pb < Ba. In terms of hydrocarbon contents, total concentrations of polynuclear aromatic hydrocarbon (PAH) compounds was 10.65 mg/kg with concentrations of the individual components (Figure 4) increasing as phenanthrene (0.78 mg/Kg: 7%) < naphthalene (1.94 mg/kg; 18%) < fluorene (2.54mg/kg; 24%) < 2-methylnapthalene (5.39 mg/kg; 51%). Results on monoaromatics (BTEX), shown in Figure 5, gave a total concentration of 4.213 mg/kg out of which toluene constituted the highest fraction (45.34%), followed by xylene (34.56%), ethylbenzene (19.63%) and benzene (0.47%). Total xylene concentration was 1.456 mg/kg out of which ortho-xylene constituted 63.46% while meta- and para-xylenes gave 36.54% of the total (1.456 mg/kg).

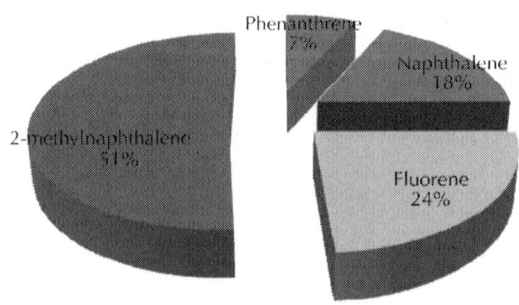

Figure 4: Percentage distribution of individual components of PAH relative to the total concentration.

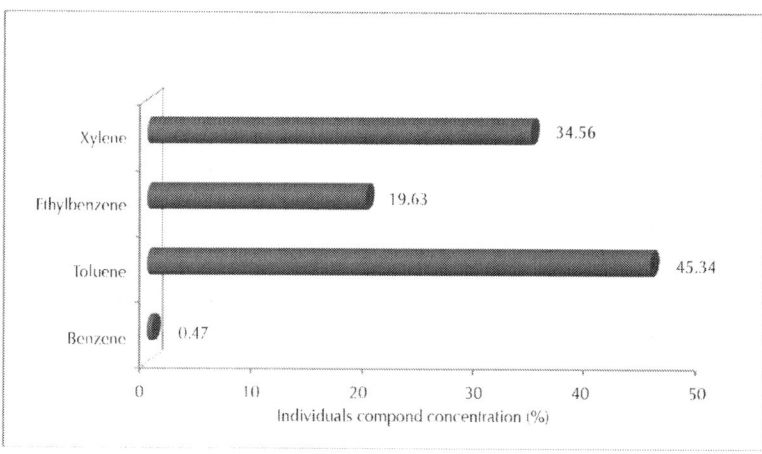

Figure 5: Percentage distribution of individual components relative to the total BTEX concentration.

Results on Petroleum Hydrocarbon Degradation

By application of CNB-Tech products, the initial TPH concentration of 79, 200 mg/kg decreased to 1888.67 ±161. 20 mg/kg. The difference in these two values was a mean TPH concentration of 77 311.33 ± 161.20 mg/kg. This difference corresponds to the total concentration of hydrocarbon compounds degraded or destroyed by the applied treatment. The initial concentration (79, 200 mg/kg) and the degraded fractions (in replicates of three) are presented in Figure 6. Specifically, results on hydrocarbon degradation (Figure 7) revealed 98% degradation for TPH, 100% degradation for BTEX and 100% degradation for PAH. Reduction in TPH level by 99% was obtained by the Nigerian laboratories.

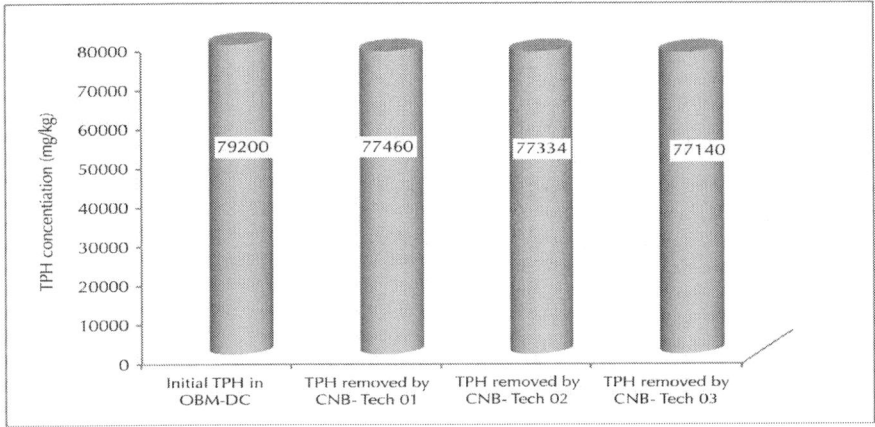

Figure 6: Graph showing concentrations of degraded TPH relative to the initial concentration.

Table 2: Qualitative results for the remediated media

S/N	Parameter	Remarks for contaminated medium	Remarks for remediated medium
1.	Appearance	Viscous, pasty and solid interfaced in oil suspension	Transformed to non-viscous, non-sticky crumby humus soil appearance
2.	Color	Light brown	Treated matrix had characteristic dark color of humus soil
3.	Odor	Presence of strong hydrocarbon odor	Complete disappearance of hydrocarbon odor in all the treated media and all treated samples exhibited clean earthy smell
4.	Sheen test	Strong oil sheen in water suspension	Complete disappearance of oil sheen in water suspension

Results on qualitative assessments of the untreated OBM-DC and remediated material in terms of appearance, odor, color and sheen test are contained in Table 2 and Figure 8 depicts the materials' appearances before and after remediation.

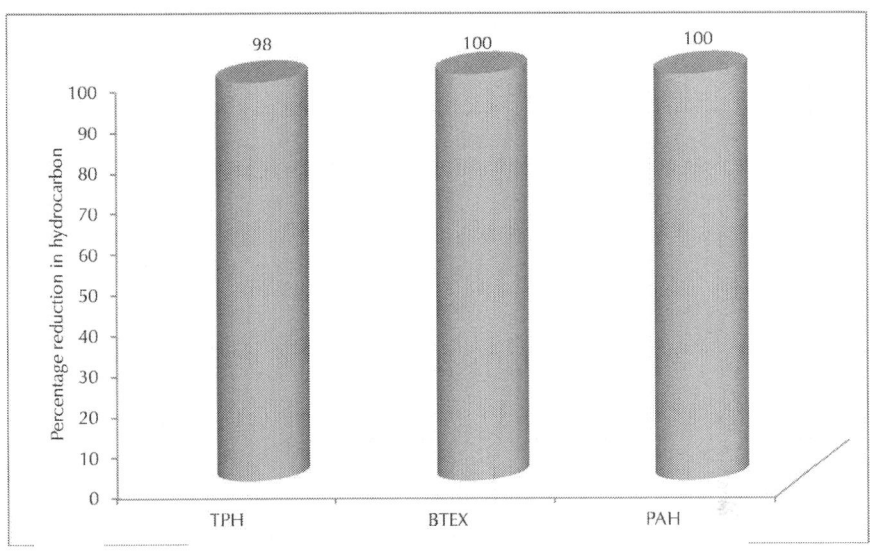

Figure 7: Percentage degradation of hydrocarbon compounds in the drilling wastes by applied CNB-Tech products.

Stabilized OBM-DC before remediation

After remediation using CNB-Tech products

Figure 8: Photographs showing the materials before and after bioremediation by the application of CNB-Tech products.

Results on Inorganic Constituents of the CNB -Tech Treated Materials

Descriptive statistics of selected inorganic constituents found in the treated media are presented in Table 3. Changes in their concentrations relative to the initial values are presented in Figure 9. For instance, the initial pH value was reduced to 7.90 from 10.20, corresponding to 23% reduction. Likewise, the following reductions were obtained: 62% for Ca, 46% for As, 44% for Cu, 70% for Pb, 100% for Hg, 57% for Ni and 37% for Zn. The concentrations of some elements such as nitrogen, phosphorus and potassium were elevated. The nitrogen-phosphorus-potassium (NPK) status, as affected by treatment, is presented in Figure 10. Nigerian laboratories obtained the same trend for NPK status. Based on the results from USA, CNB-Tech remediation option applied in this study raised the nitrogen level from 0.036% to 0.096%, raised phosphorus level from 0.0291% to 0.312%, increased potassium by 1.4 fold (Figure 10) and sodium by 3 folds. The USA based laboratory did not analyze for total organic carbon and electrical conductivity but the Nigerian based laboratory did and recorded electrical conductivity in the range of 1956 to 2063 mSm^{-1} with a mean value of 2003 ± 54 mSm^{-1} before treatment. After remediation, the electrical conductivity of the end products ranged from 594 to 696 mSm^{-1} and a mean value of 640 ± 52 mSm^{-1}. From the mean values, there was a 68% reduction in electrical conductivity.

Table 3: Concentrations of some inorganic parameters in the treated materials

S/N	Element	Minimum	Maximum	Mean	Standard error	Standard deviation	Sample population
1.	pH	7.70	8.20	7.90	0.15	0.26	3
2.	Nitrogen (%)	0.070	0.130	0.096	0.016	0.028	3

3.	Phosphorus (%)	0.280	0.360	0.312	0.026	0.046	3
4.	Potassium (%)	0.50	0.77	0.61	0.08	0.14	3
5.	Copper (mg/kg)	18.10	21.70	20.10	1.06	1.83	3
6.	Zinc (mg/kg)	79.30	110	92.67	9.08	15.73	3
7.	Nickel (mg/kg)	3.99	7.05	5.29	0.92	1.59	3
8.	Calcium (mg/kg)	28900	39200	33466	3030	5248	3
9.	Arsenic (mg/kg)	2.50	4.85	3.59	0.68	1.18	3
10.	Lead (mg/kg)	5.87	54.80	27.06	14.50	25.12	3

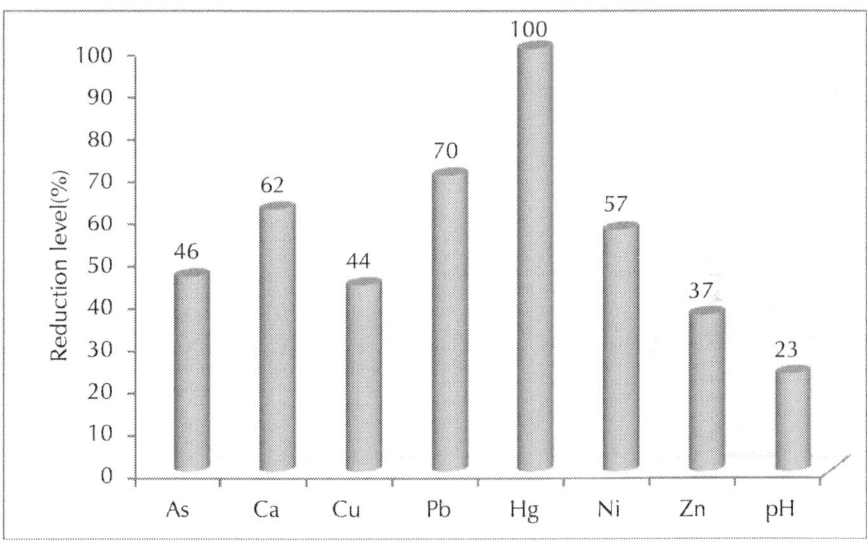

Figure 9: Reductions in some inorganic constituents of the drilling materials treated by CNB-Tech.

Total organic carbon ranged from 2.95 to 3.06% with a mean of 2.99± 0.06% before remediation and increased to 3.84 to 3.93% with a mean of 3.88 ± 0.05%; corresponding to an increase by 23%. Before remediation, Cd concentration varied from 6.70 to 7.60 mg/kg, with a mean value of 7.03± 0.49 mg/kg. After treatment, the metal concentration ranged from 0 to 1.80 mg/kg with an average of 1.05 ± 0.94 mg/kg. By the two mean values, cadmium level was reduced by 85% due to applied CNB-Tech products.

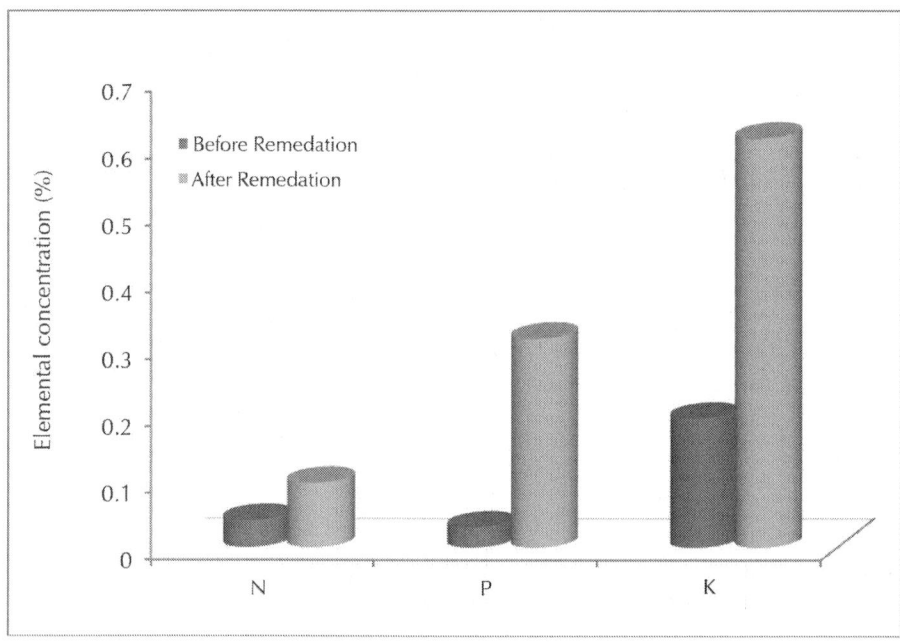

Figure 10: Nitrogen-phosphorus-potassium status before and after treatment as obtained by the USA based laboratory.

Results on Microbial Activity

The digital photographs of heterotrophic plate count results are shown in Figure 11. Microbial activities assessed on the untreated and treated samples revealed that the contaminated oil-based mud and cuttings (no. 1 in Figure 11), contained some indigenous microorganisms of up to 1.9×10^3 (cfu/mL) while the CNB-Tech remediated samples recorded up to a maximum of 3.15×10^7 cfu/mL. An illustration of microbial enumeration for 48-hr and 96 hr counts are presented in Figure 12.

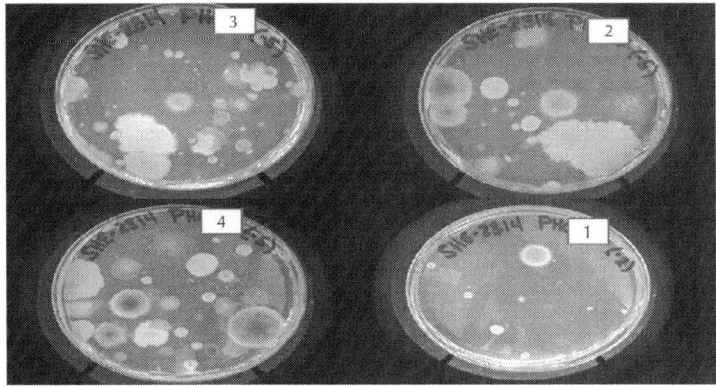

Figure 11: Heterotrophic plate count digital photographs for untreated OBM-DC (1) (before remediation) and replicates (2, 3, 4), after remediation using CNB-Tech method.

At 48 hr microbial activity assessment, maximum total microbial population of 1.9×10^3 cfu/mL was obtained for untreated OBM-DC and in the materials remediated by the application of CNB-Tech products, it was 1.45×10^7 cfu/mL. These two values were significantly different at $p \leq 0.05$. At 96 hr microbial activity assessment, a total microbial population of 2.4×10^3 cfu/mL was obtained for untreated OBM-DC and 3.15×10^7 cfu/mL for the remediated matrices. Results showed that within 48 hours, the microbial activity of the remediated matrices excelled over the untreated by over 7,000 folds and at 96 hours, it excelled by over 13, 000 folds, indicating rapid multiplication of microbial activity by CNB-Tech products which also increased with time.

Results on Phytotoxicity Assessment of Remediated Samples

Toxicity on Seed Germination Potential

The contaminated OBM-DC did not allow the germination of maize seedlings. Out of the sown 18 seedlings, none germinated.

The untreated OBM-DC therefore, gave 100% toxicity to seed germination potential (SGP) of maize. On the contrary, all the 18 maize seedlings sown in the CNB-Tech remediated matrices germinated (Figure 13). Hence, resulting in 100% positive effect on SGP, indicating that the treated matrices exhibited 0% toxicity to seed germination.

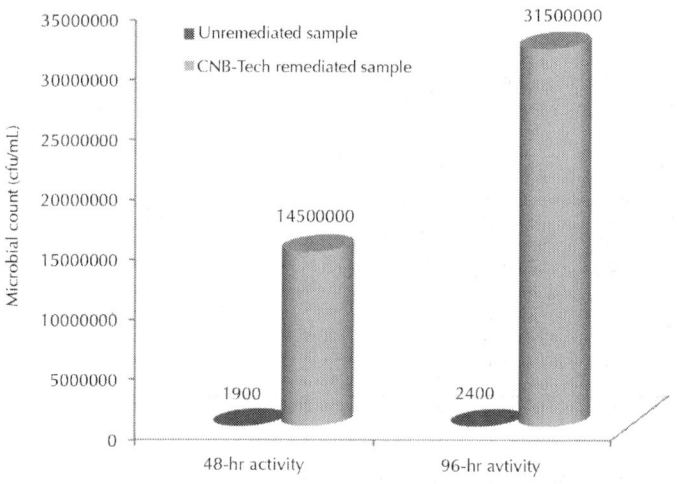

Figure 12: Microbial activity at 48 –hr and 96-hr counts for untreated oil-based drilling wastes and CNB-Tech remediated samples.

Figure 13: Germinated maize seedlings growing in treated media with picture taken on day 4 of growth.

Results on Beneficial Use of Remediation End Product

Figure 14, shows a cross-section of the treated materials (during recovery period) being aerated in preparation for use as plant growth media.

Figure 14: A cross section of project technical staff preparing the treated drilling wastes (OBM-DC) for use as plant growth media.

During the recovery phase of the remediated end-product, treated materials were allowed to lie fallow in order to establish natural processes as a sign of wellbeing and restoration. In this project, after the fallow period, early indications of material restoration were:

- spontaneous vegetative growth,
- the presence of larva within the spontaneously grown green vegetation,
- butterflies and small birds perching on the surface of the material, which could not take place before treatment

Remediated materials supported the growth of fluted pumpkin (*Telfairia occidentalis*). A cross-section of the green leafy vegetable at over 100 days of growth and that of cassava, at one week of growth, growing in the treated materials are shown in Figure 15. Narrowing to the height of *Telfairia occidentalis*, the mean height for crops grown in the untreated OBM-DC was 0 cm as there was complete inhibition to both germination and growth. The mean height for crops grown in CNB-Tech remediated media was 217± 25 cm, a value higher than the mean height (187± 40 cm) of the vegetable crops grown in farm soil collected from the region. The difference in the two mean values was significant at $p = 0.14$. Correlation for the heights of the vegetables grown in the treated media and those grown in the farm soil gave a coefficient of 0.95 ($p = 0.204$).

Fluted Pumpkin

Cassava

Figure 15: Remediated drilling wastes as plant growth medium for Fluted pumpkin (*Telfairia occidentalis*) and cassava (*Manihot esculenta Crantz*).

Results on the Impact of Remediation Leachate on Plant Life

Comparative evaluations of control system (soil treated with water only), stock leachate system (soil treated with leachate without any form of dilution) and systems treated with serial dilutions of the leachate (soil treated with leachate diluted with water by factors 1, 2, 3 and 4) are presented in Table 4.

Table 4: Impact of leachate generated at the close-out of project on the root length and height of maize

S/N	System Code	Leachate effect of on vegetative growth relative to control (%)		Effect of serial dilution on plant using stock (undiluted leachate) as reference (%)	
		Height	Root length	Height	Root length
1.	FS+ Water (Control)	Reference	Reference	Not applicable	Not applicable
2.	FS + DF-0	-1.50	-23.45	Reference	Reference
3.	FS + DF-1	32.60	1.12	34.62	32.20
4.	FS + DF-2	45.01	16.37	42.22	50.02
5.	FS + DF-3	66.86	21.37	69.41	58.55
6.	FS + DF- 4	75.39	24.51	78.07	62.66

[i] - Negative sign stands for decrease. The other positive values stand for increase, FS = farm soil and DF = dilution factor

Pictorial and graphical representations of leachate impact on plant height and root length are presented in Figures 16 and 17. Relative to the control system (soil treated with water only), leachate diluted with water by a factor of 4 improved plant height by 75.39% and root length by 24.51%. Figures16 and 17gave all the systems at a glance, relating the control (FS + Water), system SF+LDF-0 (DF-0) and serial dilutions (DF-1 = FS+ LDF-1, DF-2 = FS+ LDF-2, DF-3 = FS+ LDF-3 and DF-4 = FS + LDF - 4) for plant height and root length. Evaluating the effect of leachate dilution relative to the stock (undiluted) leachate, a 4-fold dilution excelled over the stock by 78.0% for plant height and 62.66% for root length. The relationships between plant height or root length and dilution factors are given in Figure 18. Pearson correlations gave strong coefficients: plant height versus dilution factor, $r = 0.979$ (p = 0.004), root length versus dilution factor, $r = 0.932$ (p = 0.021) and plant height versus root length, $r = 0.972$ (p = 0.006). From the results, plant vegetative growth increased with increasing dilution of leachate.

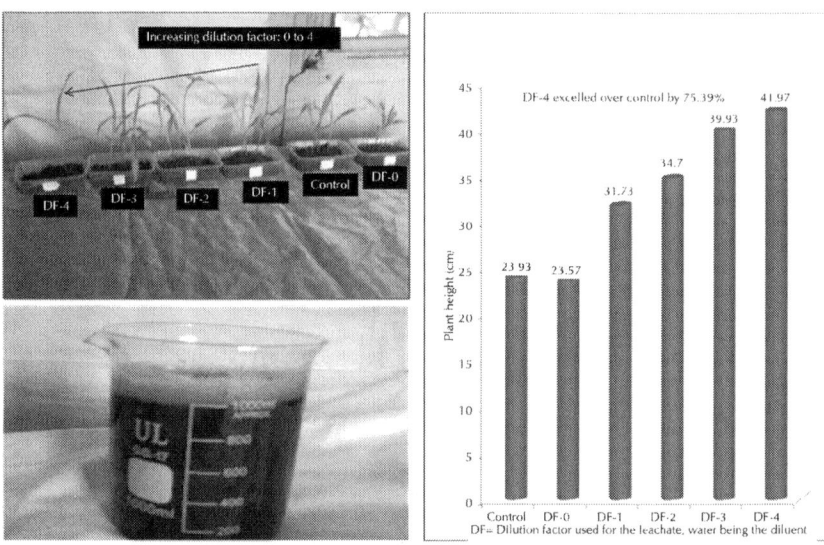

Figure 16: Pictorial and graphical representations of leachate impact on height of maize, including a picture of the stock leachate contained in a beaker.

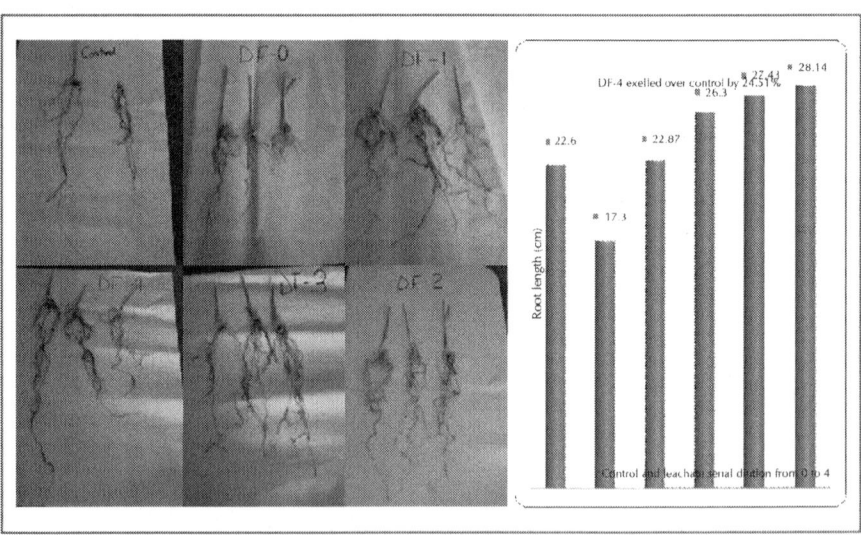

Figure 17: Pictorial and graphical representations of leachate impact on root length of maize.

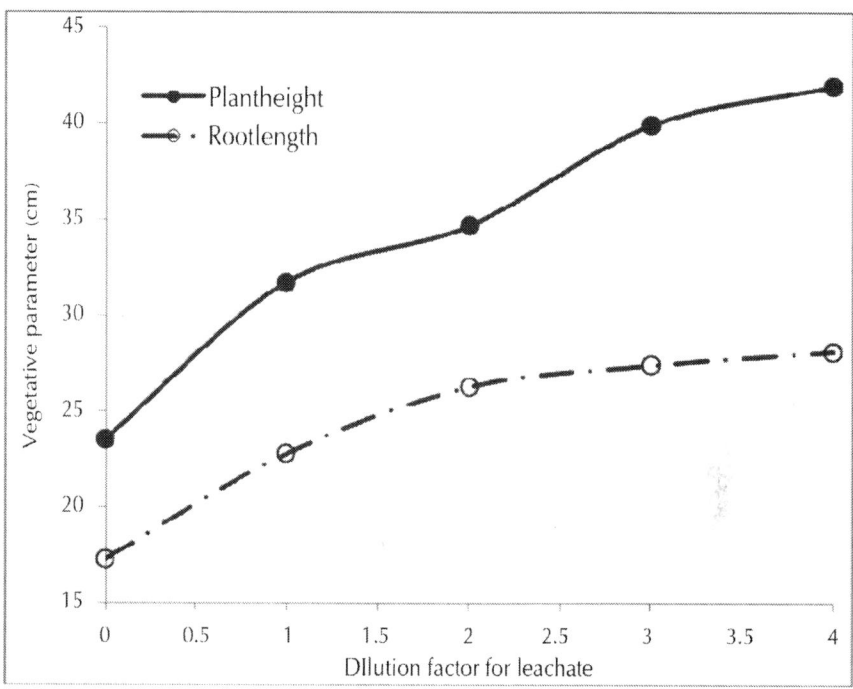

Figure 18: Relationship between plant vegetative growth and serial dilution of process fluid (leachate) generated during the remediation project.

DISCUSSION

The type of inorganic constituents and hydrocarbons found in the drilling wasting used in this study were consistent with the reports of [4, 17] but varied in concentrations. This confirms that the OBM-DC used in this study was toxic [2]. The remediation products of CNB-Tech series used in this study demonstrated a high (98 to 100%) degradation potential for the different constituents of hydrocarbon compounds found in the drilling wastes, within a short period of 6 days. This excellent performance was attributed to the chemistry, nature and operation mechanisms of the CNB-Tech formulations.

An infrared spectrum is primarily used to identify functional groups present in a molecular fragment [33]. The infrared spectra obtained for CNB-Tech products (Biozator and Bioprimer) revealed enrichment of the molecular structure of the two products with oxo- groups, indicating oxidizing functionality. The presence of C-H of aromatic nature and the O-H stretching absorption indicate the presence of both hydrophobic and hydrophilic properties, respectively, in their molecular fragments. By implication, the remediation products are naturally endowed with:

- oxidizing ability
- polar (hydrophilic: water loving) molecular fragment
- non-polar fragment (hydrophobic: water insoluble, oil soluble) molecular fragment.

These natural endowments permit the dissolution of the products' active ingredients (solids) in water, making water the carrier medium for CNB-Tech liquid formulations. Consequently, Biozator and bioprimer are water based technical grade products. By the mentioned characteristics, the two products perform reduction and oxidation (Redox) reaction mechanisms, resulting in the degradation/ destruction of hydrocarbons compounds, without recombination to form new hydrocarbons. These absorption peaks in the infrared spectra further reveal that CNB – Tech products are natural hydrocarbon biodegradation catalysts for the following reasons:

- enhaced water solubility of hydrocarbons via sorption, hydrolysis and oxidation mechanisms
- enhanced bioavailability of hydrocarbon pollutants for microbial degradation
- increased supply of oxygen [O] molecules required for enhanced reduction –oxidation reactions in the hydrocarbon degradation process.
- surfactant property
- emulsification of hydrocarbons

The combined actions of hydrophobic molecular fragment, hydrolysis, oxidation and surfactant property of CNB-Tech products render hydrocarbons more water soluble and subsequently more available for biodegradation. Bioprimer and Biozator also emulsify hydrocarbons into droplets that can be easily assimilated by microorganisms. By these properties, the products reduce oil-water surface tension; enhance water solubility of petroleum hydrocarbons thereby enhancing the bioavailability of the contaminants (hydrocarbons) to microorganisms for both extracellular and intracellular decompositions. The two products are 100% biodegradable. The third CNB-Tech product used in this study (Ecorem: a black amorphous solid material, also 100% biodegradable) contains major and minor plant nutrient elements and via hydro-activation, naturally generates mixed consortia of microorganisms, which multiplies with time to facilitate the destruction of hydrocarbons. No engineered microorganism or externally imported microorganism was used in this study. This technology, therefore, saves time and eliminates the daunting task of isolating pure microbial strains and associated adaptability challenges linked with conventional bioremediation techniques [7, 8, 18, 19, 20].

The microorganisms from Ecorem product perform the following functions:

- extracellular decomposition in which the naturally produced microorganisms secrete enzymes to breakdown large organic compounds (such as hydrocarbons) into smaller forms for easier absorption into the micro-organisms. Once the smaller compounds have been absorbed by the microorganisms, intracellular decomposition takes place

- increased microbial activity facilitated by Ecorem, results in thermophilic temperature modulations in the range of 55 to 60°C, a process that accelerates degradation of hydrocarbons, especially polynuclear aromatic aromatic hydrocarbons (PAHs). Thermophilic temperature modulations also controls thermo-sensitive pathogen to crops animals and man; killing off weeds and seeds that will be detrimental to land use of end products.

By the above described mechanisms, the CNB-Tech products were able to biodegrade petroleum hydrocarbon compounds with high efficiency (98% degradation for TPH and 100% degradation for PAHs and BTEX) within a short period of time of 6 days, relative to previous works on bioremediation. For instance, in a study of in-situ bioremediation of oily sludge via biostimultaion of indigenous microbes, conducted by [34], through the addition of manure at the Shengli oilfield in Northern China for 360 days, 58.2% reduction in TPH was achieved in test plots and 15.5% reduction in control plot. By treating 2 kg of drill cuttings with initial TPH of 806.36 mg/kg for 56 days under the conditions of composting of spent oyster mushroom (*P.ostreatus*) substrate, [35] recorded overall degradation of PAHs in the range of 80.25 to 92.38%. In this present study, OBM-DC used had initial TPH of 79, 200 mg/kg and was degraded by 98% within the stated short period of 6 days. In a field trial biopile composting method [36] for drilling mud polluted sites in the Southeast of Mexico with comparable TPH level of 99 300 ± 23000 mg/kg, after 180 days, TPH concentrations decreased from 99 300 ± 23000 mg/Kg to 5500 ± 700 mg/kg, corresponding to 94% degradation for amended biopile and to 22900 ±7800 mg/kg, representing 77% decrease for unamended biopile. The mean residual value of TPH (5500 ± 700 mg/kg) left in the treated matrix in their study was higher than the mean residual value (1888± 161 mg/kg) obtained in this present study.

By conducting an investigation on two bioremediation technologies (bioremediation by augmentation and conventional composting using crude manure and straw) as treatment options for oily sludge and oil polluted soil in China [12] in which the total hydrocarbon content (THC) varied from 327.7 to 371.2 g/kg (327700 to 371200 mg/kg) for dry sludge and 151.0 g/kg (151000 mg/kg) for soil for a period of 56 days; after three times of bio-preparation application, THC decreased by 46 to 53% in the oily sludge and soil. The results (98 -100% degradation) obtained from this present study was from only one dose application of CNB-Tech products. Repeated application of CNB-Tech products by two to three dose applications will achieve 100% degradation of

TPH. In another instance, a 5- month field scale bioremediation of sludge matrix via the utilization of organic matter such as bark chips via conventional composting, mineral oil (equivalent to total hydrocarbons) decreased from 2400 to 700 mg/kg (70% decrease) for sludge matrix and from 700 to 200 mg/kg, corresponding to 71% decrease [14]. In treating oil sludge using composting technology in semiarid conditions for 3 months, hydrocarbons were reduced from 250 to 300g/kg (250000 to 300 000 m/kg) by 60% against reduction by 32% recorded in the control [37]. The treatment applied by [37] and consequent reduction of 60% implies that the residual hydrocarbons in the treated samples would be between 100 000 and 180 000 mg/kg unlike the results obtained in this present study that gave residual hydrocarbon of 1888.67 ±161.20 mg/kg. In a study carried out by [38], sand samples contaminated with oil spill were collected from Pensacola beach (Gulf of Mexico) and tested to isolate fungal diversity associated with beach sands and investigate the ability of isolated fungi for crude oil biodegradation. From their results, 4.7 to 7.9% biodegradation was recorded.

Elsewhere in India, Abu Dhabi and Kuwait [39], bioremediation technology was applied in field-scale degradation of hydrocarbons in different oil wastes for a period of 12 months. Table 5 illustrates different reductions in total petroleum hydrocarbons obtained in these field case studies. TPH reductions in drilling wastes were obtained in the range of 90.85 to 95.48% with residual TPH in treated samples in the range of 2600 to 10 900 mg/kg (0.26 to 1.09%).

Table 5: Reductions in TPH levels obtained in field case studies of different types of petroleum impacted wastes (soils, drill cuttings and oil-based mud) in Abu Dhabi, Kuwait and India [39]

Name of the oil Installation / type of oily waste	Quantity of oily waste (cubic meter)	Number of batches	TPH Content (%) in oily waste before and after bioremediation		% Reduction in TPH	Residual TPH in treated material (%)
			Before	After		

Abu Dhabi National Oil Company (ADNOC), Abu Dhabi / Oil contaminated drill cuttings	200	1	17.26	0.98	94.32	0.98
BG Exploration and Production India Limited (BGEPIL), India / Oil based mud (OBM)	2,428	3	5.75 – 6.23	0.26 - 0.57	95.48-90.85	0.26 – 0.57
Bharat Petroleum Corporation Limited (BPCL), India / Oily sludge	5,000	1	19. 30 – 26.5	0.26 - 0.57	98.65-97.85	0.26 -0.57
Cairn Energy Pty. India Limited, India / Oil contaminated drill cuttings	567	2	14.93 – 18.81	0.82 – 1.09	94.51-94.21	1.09
Chennai Petroleum Corporation Limited (CPCL), India / Oily sludge	4,444	2	26.12	0.89	96.59	0.89
Hindustan Petroleum Corporation Limited (HPCL), India / Oily sludge	5,010	3	16.70 – 52.81	0.90 – 1.60	94.61-96.97	0.90-1.60
Indian Oil Corporation Limited (IOCL) Refineries in India / Oily sludge (acidic + non acidic)	75,412	48	9.6 – 38.4	0.37 – 0.95	96.15-97.53	0.37-0.95
Kuwait Oil Company (KOC), Kuwait / Oil contaminated soil	778	1	4.6 – 12.75	0.09 – 0.10	98.04-99.21	0.09-0.10

Mangalore Refinery and Petrochemicals Limited (MRPL), India. / Oily sludge	2,222	2	8.35 – 19.86	0.84 – 0.97	89.84-95.12	0.84-0.97
Oil and Natural Gas Corporation Limited (ONGC) installations in India / Oily sludge & oil contaminated soil	95,499	145	12.0 – 51.5	0.5 – 1.2	95.83-97.67	0.50--1.20
Oil India Limited (OIL), Assam / Oily sludge & oil contaminated soil	15,921	14	21.6 – 37.7	0.49 – 0.53	97.73-98.59	0.49-0.53
Reliance Energy Limited (RIL), India / Oily sludge	611	2	19.15	0.5	97.39	0.50

The residual TPH level (1888.67 ± 161.20 mg/kg) obtained in this present study was below the Environmental Guidelines and standards for the Petroleum Industry in Nigeria (EGASPIN) intervention value for mineral oil (petroleum hydrocarbon) of 5000 mg/kg [15]. By repeated application of CNB-Tech products, it is possible to meet a very strict regulatory standard for residual TPH level of less than 50 mg/kg. The changes in metal concentrations found in this study were attributed to (i) immobilization via chelate formation (ii) preferential supplementation of trace plant nutrient elements using the three products, (iii) natural electrochemical process whereby the positively or negatively charged organic molecules (generated during the natural transformation process occurring when the products were in use) bond with their counterparts in organic matter. These processes include oxidation, methylation, hydroxylation, carboxylation, coupling and polymerization [40] thereby enhancing bioavailability of the metals to microorganisms that utilize the organic matter supplied by the CNB –Tech products as energy source. Microbial population found in a typical tropical soil under Nigerian climate is in the neighborhood of 8.19 x 10^6 cfu/mL [41]. Relative to this value, the population found in the contaminated OBM-DC (1.9 to 2.4 x 10^3 cfu/mL) showed suppressed microbial population, attributed to strong hydrocarbon (TPH level of 79, 200mg/kg) pollution. This is in agreement with the reports of [3]. The microbial population (1.45 to 3.15 x 10^7 cfu/mL) found in treated samples revealed restoration of soil microbial population using CNB-Tech products. It excelled over the value recorded in polluted material by over 7000 folds and higher than the value reported by [34], where TPH degraders and PAH degraders increased by one to two orders of magnitude via the addition of manure. Furthermore, the use of CNB-Tech products modified the pH value of the drilling wastes, transforming it from strongly alkaline (pH of 10) medium to pH of 7.90 medium; comparable to the 7.3±0.1 obtained by [34] for bioremediated soils. The very high pH of the untreated drilling waste materials could be attributed to some of the additives in the drilling fluid. Drilling fluids contain an internal phase of brine such as calcium salts [3]. This was confirmed by the high content of Ca (87 300 mg/kg) obtained in this study for the untreated material. One dose application of CNB-Tech products reduced this concentration by up to 62%, repeated dose application would definitely bring Ca level to any desired value.

Observations made during the recovery /fallow period were signs of drastic positive change in toxicity conditions, implying reduced toxicity. Reduction of soil toxicity by bioremediation, evidenced by increase in EC50 of the soil was reported by [34]. In this study, bioremediation using CNB-Tech products reduced toxicity in treated materials relative to untreated OBM-DC, evidenced by 100% positive effect on seedling germination potential and improved crop vegetative growth. Reduced material toxicity also explains the increased microbial activity of the treated matrices in comparison to the untreated drilling wastes, obtained in this study. The agricultural potential for the remediation end-products was also manifested by:

- increased microbial activities
- increased nitrogen-phosphorus-potassium (NPK) status
- increased soil crumby nature as against very viscous and pasty characteristics of untreated drilling wastes.

These nutrient elements (NPK) enhance microbial growth, microbial population, microbial activity and consequently increase soil fertility [41]. By these, CNB-Tech products could overcome the extreme phytotoxicity [100% toxicity to seedling germination potential of maize and 100% inhibition to vegetative growth for three different types of plant (maize, fluted pumpkin and cassava)], caused by the untreated drilling waste. CNB-Tech products transformed oil-based drilling mud/cuttings to arable soil; capable of supporting seed germination and plant growth; excelling the performance of a control (farm soil apparently not impacted by drilling waste or crude oil) by 14%.

Electrical conductivity, a measure of dissolved ions in solution, is influenced by several soil physical and chemical properties such as salinity, saturation percentage, water content, bulk density, organic matter content, temperature and cation exchange capacity of the soil matrix. Impact of these influencing factors must be reflected in interpreting electrical conductivity effect on plant growth. Generally, elevated electrical conductivity and high salinity levels in agricultural soils may result in reduced plant growth and productivity or in extreme cases, the elimination of crops and native vegetation [42]. The reduction of electrical conductivity by 68% is a positive development because it demonstrates that the products could also modify the salinity of the material. In situations of very high initial electrical conductivity, there is a step-down CNB-Tech product as was carried out in this study and in situations of very low electrical conductivity, there is also a step-up CNB-Tech product as reported in a previous publication [30].Results in this present study on excellent growth of crops planted in the remediated matrices were indicators of acceptable soil salinity level for plant growth. The beneficial use of the end-products obtained in this study for crop production were attributed to postulations based on findings from this study and previous works on this subject matter, which include:

- stimulation of beneficial microorganisms in soil, which enhances soil fertility [25]
- possible increased photosynthetic rate in plants evidenced by increased photosynthetic pigments (chlorophylls a and b) [40]
- increase in soil buffering capacity [28]
- increased soil moisture retention capacity by reducing hydrophobicity tendency [29]
- positive soil temperature modifications that enhance soil nutrient bioavailability to plants [31, 40]
- formation of stable chelates with toxic metals such as Pb, Cu and Cd in order to reduce their bioavailability to plants [40]
- preferential exclusion of the chelated toxic metals from soil solution, allowing the plant nutrient elements to be assimilated into plant cells
- improvement of soil physicochemical properties via:
- increased aeration and water retention [29]
- activation of the macro and micro nutrients in soil in forms readily assimilated by plants [30, 40]
- improvement of plant root development and growth
- improvement of seed sprout of plants and subsequent shoot growth
- improved plant biomass production [26]
- enhanced soil nitrogen, phosphorus and potassium status for improved soil fertility
- acting as plant growth hormone, having positive stimulant action for plant growth [25, 26]
- improvement of soil permeability, promoting plant drought resistance [29]
- promotion of increased soil porosity and organic matter content, hence greatly promoting the microorganism activity and improving soil fertility.

Regarding leachate generation and management during the remediation exercise; fluid (leachate) produced as remediation progressed was recycled by incorporation into the biocell and used to regulate moisture content, thereby reducing water usage and conserving water resources. Expertise applied during the project ensured that at remediation project close-out, no isolated fluid system was actually produced. Nonetheless, the assessment of leachate effect on plant growth carried out in this work was to establish the fact that even in the event of accidental release of some fluid into the environment, there would be minimal risk to the receptor biotic community. More evaluations are still ongoing in this regard. Results from this study revealed that the leachate generated, though a concentrate, supported plant growth and when diluted with ordinary tap water gave a better support; reasons being that:

- toxic petroleum hydrocarbons in the contaminated drilling wastes have been destroyed to an acceptable level, evidenced by natural foamability of the concentrated leachate. Foamability would hardly occur if oil was still present
- leachate is also enriched with plant nutrients such as nitrogen, phosphorus and potassium

The process fluid, therefore, had some fertilizer value. The percentage decreases (1.50% and 23.45%) obtained for plant height and root length respectively, for the stock leachate was attributed to concentrated level of nutrients, confirmed by better performance of dilute leachate series. Naturally, in any formulated fertilizer, plant nutrients are applied at specified concentrations otherwise may hinder plant growth. Comparative evaluations of control system (soil treated with water only), stock leachate system (soil treated with leachate without any form of dilution) and systems with serial dilutions of the leachate (soil treated with leachate diluted with water by factors 1, 2, 3 and 4) revealed that the leachates were not toxic to receptor plants. The implication of this is that in the event of occasional spill of the leachate to the adjacent environment; dilution with water is, therefore, an adequate safety measure.

The ability of the end products to sustain the growth of green leafy vegetable: fluted pumpkin (*Telfairia ocidentis*) and root tuber crop, cassava (*Manihot esculenta Crantz*) and cereal crop (maize) is a demonstration of the utility of the remediation end product. It therefore stands that the use of CNB-Tech products as a biotechnological tool for hydrocarbon degradation in drilling waste converts these waste materials into non-toxic and potentially useful end products. In addition to the beneficial use of the remediation end-product for agricultural purposes, other possible utility options, shown in Figure 19, include:

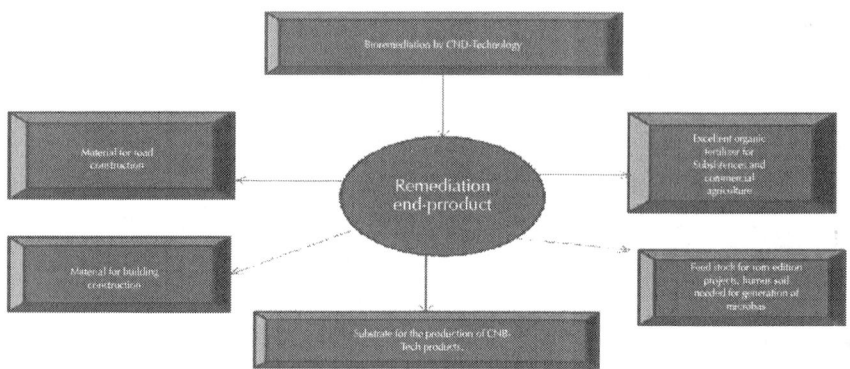

Figure 19: Potential utility of end - products from bioremediation using CNB-Tech products.

- material for road construction
- material for building construction
- substrate for the production CNB-Tech bioremediation agents
- excellent organic fertilizer for subsistence and commercial agriculture
- feedstock for bioremediation projects

Table 6 is a comparative evaluation of economic, operational and environmental implications of thermal technologies as reported by [3] and CNB-Technology based on the results and learning from this study.

Table 6: Comparative evaluation between thermal technology and CNB-technology

S/N	Thermal Technology	CNB-Tech
1.	Effective removal and recovery of hydrocarbons from solids	Effective removal of hydrocarbons from solid
2.	Possibility of recovering base fluid and end - product could be used for brick making	Effective recovery of free phase oil and end product has other uses apart from brick making
3.	Low potential for future liability	No future liability
4.	Requires short time	Time is relatively short
5.	High cost of handling environmental issues, since end- product dispersion would be below organic layer where vegetation growth is desired	Very minimized environmental issues
6.	Large volume of wastes is required to justify the cost of operation	Cost-effective for either small or large volume of wastes
7.	Requires tightly controlled process parameters	Does not require tightly controlled process parameters
8.	Heavy metals and salts are concentrated in processed solids	Reduces heavy metals and salts concentrations in process solid
9.	High operating temperatures can lead to safety risks	Low operating temperature. Operates at ambient temperature; modulation does not exceed 60°C.
10.	Requires several operators	Does not require several operators
11.	Process water contains some emulsified oils	Process water does not contain some emulsified oils
12.	Residue ash requires further treatment	No residue ash. End-product is clean soil
13.	End product is sterile and can no longer support plant Life	End product is fertile and can support microbial and plant Life

CONCLUSIONS AND RECOMMENDATIONS

This study revealed that it is possible to harness natural, biodegradable and local resource materials of Nigeria origin; translate them to scientifically formulated products that can be utilized for efficient biodegradation of hydrocarbon polluted matrices such as oil-based mud and drill cuttings within a reasonable short period of 6 days. This technology thus converts hydrocarbon polluted oil-based mud and drill cuttings to beneficial end-products of high order reuse such as soil amendment, without the generation of secondary waste materials. Field-scale trial adopting CNB-Technology is recommended.

ACKNOWLEDGMENTS

This project was carried out under full financial support of the Remediation Department, Shell Petroleum Development Company (SPDC), Port Harcourt, Nigeria through the University Liaison Team of the company. The support of the Oil well Team of SPDC that facilitated the procurement of oil- based mud and drill cuttings is also acknowledged.

REFERENCES

1.   United Nations Environmental Programme (UNEP), 2011. Environmental Assessment of Ogoniland. P.1-262. ISBN:978-92-807-9 Available on line at: http://postconflict.unep.ch/publications/OEA /UNEP_OEA.pdf

2.   Department of Health, Government of South Australia (DHGSA). Public Health Fact Sheet on Polycyclic Aromatic Hydrocarbons (PAHs): Health effects 2009 http://www.dh.sa.gov.au/pehs/PDF-files/ph-factsheet-PAHs-health.pdf

3.   Neff, M.M and Duxbury, MA. Composition, environmental fates, and biological effects of water based drilling muds and cuttings discharged to the marine environment: A Synthesis and Annotated Bibliography. Prepared for Petroleum Environmental Research Forum (PERF) and American Petroleum Institute. 2005. http://perf.org/pdf/APIPERFreport.pdf

4. Gbadebo, A.M., Taiwo, A.M. and U. Eghele, U Environmental impacts of drilling mud and cutting wastes from the Igbokoda onshore oil wells, Southwestern Nigeria. Indian Journal of Science and Technology, 2010; 3(5), 504 -510.

5. Environmental Protection Agency (EPA). An Assessment of the Environmental Implications of Oil and Gas Production: A Regional Case Study, 2008

6. Osuji, L.C., Erondu, E.S and Ogali, R.E Upstream petroleum degradation of mangroves and intertidal shores: The Niger Delta Experience. Chemistry and Biodiversity, 2010: 7, 116 -128.

7. Knez, D., Jerzy, A, G and Czekaj Trends in the drilling waste management. Acts Montanistica Rocnlk, 2006:11, 80-83.

8. Morillon, A., Vidalie, J.F., hamzah, U.S., Suripno and Hadinota, E.K "Drilling and Waste management", SPE 73931, Intenationa; Conference on Health, Safety and Environment in oil and gas exploration and production, 2002: March 20-22

9. Zimmerman, P.K. and Rober, J.D Oil-based drill cuttings treated by landfarming. Oil and Gas J, 1991: 12, 81-84

10. Rojas-Avelizapa, N.G., Roldan-carrillo, T., Zegarra-Martinez, H., Munez-Colunga, A.M and Fernandez-Linares A field trial for an ext-situ bioremediation of a drilling mud-polluted site. Chemosphere 2007: 66, 1595-1600.

11. Frydda, S and Randle, J.B Case study: Biological treatement of Geothermal drilling cuttings. Proceedings World Geothermal Congress, Bali, Indonesia, 25-29, 2010: 1-3.

12. Ouyang, W., Liu, H., Murygina, V., Yu, Y., Xiu, Z and Kalyuzhnyi, S Comparison of bio-augmentation and composting for remediation of oily sludge: A field-scale study in China. Process Biochemistry, 2005: 40, 3763 -3768.

13. Vidali, M. Bioremediation: An overview. Pure and Applied Chemistry, 2001: 73(7), 1163-1173

14. Jorgensen, K.S., Puutstinen, J and Suortt, A. –M Bioremediation of petroleum hydrocarbon-contaminated soil by composting

in biopiles. Environmental Pollution, 2000: 107, 245-254.

15. Department of Petroleum Resources. Environmental Guidelines and Standard for the Petroleum Industry in Nigeria, 2002

16. Joel, O.F and Amajuoyi, C.A Determination of selected physicochemical parameters and heavy metals in a drilling cutting dump site at Ezeogwu–Owaza, Nigeria. J. Appl. Sci. Environ. Manage, 2009: 13(2), 27- 31.

17. Okparanma, R.N., Ayotamuno, J. M Polycyclic aromatic hydrocarbons in Nigerian oil-based drill-cuttings; evidence of petrogenic and pyrogenic effects. World Applied Sciences Journal 2010; 11 (4): 394-400, ISSN 1818-4952.

18. Nweke, C.O and Okpokwasili, G. C Drilling fluid base oil biodegradation potential of a soil Staphylococcus species. African Journal of Biotechnology 2003; 2 (9), pp. 293-295. http://www.academicjournals.org/AJB

19. Ayotamuno, J.M., Okparanma, R, N and Araka, P.P Bioaugmentation and composting of oil-field drill-cuttings containing polycyclic aromatic hydrocarbons (PAHs). Journal of Food, Agriculture & Environment 2009; l.7 (2): 6 5 8 - 664. www.world-food.net

20. Okparanma, R.N Ayotamuno, J.M and Araka, P.P Bioremediation of hydrocarbon contaminated-oil field drill-cuttings with bacterial isolates. African Journal of Environmental Science and Technology 2009 3 (5), pp. 131-140. Available online at http://www.academicjournals.org/AJEST

21. Ifeadi, C.N The treatment of drill cuttings using dispersion by chemical reaction (DCR). A paper prepared for presentation at the DPR Health, Safety & Environment (HSE) International Conference on Oil and Gas Industry in Port Harcourt, Nigeria. 2004.

22. Adekunle, I.M., Ajijo, M.R., Omoniyi, I.T and Adeofun, C.O Response of four phytoplankton species in some sections of Nigeria coastal waters to crude oil in controlled ecosystem.

Int. J. Environ., Res., Iran, 2009; 4 (1): 65 -74 http://ijer.ut.ac.ir

23. Adekunle, I.M and Onianwa, P.C Functional group characteristics of humic acid and fulvic acid extracted from some agricultural wastes. Nigerian Journal of Science, Nigeria, 2001: 35 (1), 15 – 19.

24. Adekunle, I.M Evaluating environmental impact from utilization of bulk composted wastes of Nigerian origin using laboratory extraction test. Environmental Engineering and Management Journal 2010; 9 (5): 721 -729.: http://omicron.ch.tuiasi.ro/EEMJ/

25. Adekunle I.M., Adekunle, A.A., Akintokun, A.K., Akintokun, P and Arowolo,T.A Recycling of organic wastes through composting for land applications: a Nigerian experience. Waste Management & Research 2011; 29 (6): 582 – 593. DOI: 10.1177/ http://wmr.sagepub.com/content/29/6/582.abstract

26. Adekunle, I.M Bioremediation of soils contaminated with Nigerian petroleum products using composted municipal wastes. Bioremediation Journal, 2011; 15 (4): 230-241, DOI: 10.1080/10889868.2011.624137. http://dx.doi.org/10.1080/10889868.2011.624137

27. Adekunle I.M., Oguns, O., Shekwolo, P.D., Igbuku, O.O and Ogunkoya, O.O Assessment of population perception impact on value-added solid waste disposal in developing countries, a case study of Port Harcourt City, Nigeria. In: Xiao-Ying, Y (Ed) Municipal and Industrial Waste Disposal. Intech; 2012, p177-206.

28. Adekunle A. A., Adekunle, I.M., Igba, T. O Assessing the effect of bioremediation agent from local resource materials in Nigeria on soil pH. Journal of Emerging Trends in Engineering and Applied Sciences 2012; 3 (3) 526-532. http://jeteas.scholarlinkresearch.org/articles/Assessing%20the%20Effect%20of%20Bioremediation%20Agent.pdf

29. Adekunle A.A., I.M. Adekunle and Igba, T.O Impact of bioremediation formulation from Nigeria local resource

materials on moisture contents for soils contaminated with petroleum products. International Journal of Engineering Research and Development 2012; 2(4) 40-45 http://www. ijerd.com/paper/vol2-issue4/F02044045.pdf

30. Adekunle A.A, Adekunle, I.M. and Igba, T.O Assessing and forecasting the impact of bioremediation product derived from Nigeria local raw materials on electrical conductivity of soils contaminated with petroleum products. Journal of Applied technology in Environmental Sanitation 2012; 2 (1) 57 -66. http://www.trisanita.org/jates/atespaper2012/ ates09v2n1y2012.pdf

31. Adekunle A.A., I. M. Adekunle and Igba T. O Soil temperature dynamics during bioremediation of petroleum products using remediation agent for Nigerian local resource materials. International Journal of Engineering Science and Technology 2012; 1 (4): 1-8. http://www.ijert.org/browse/june-2012-edition

32. Association of Official Analytical Chemists (AOAC), Official Method and Analysis of The Association oh The Official Analytical Chemists 11th Edition Washington D C, 1970.

33. Finar, I.L Organic Chemistry, volume I The Fundamental principles. 6th Ed, Longman, 1973.

34. Liu, W., Luo, Y and Teng, Y Bioremediation of oily sludge-contaminated soil by stimulating indigenous microbes. Environ Geochem health 2010: 32, 23 -29.

35. Ayotamuno, J.M., Okparanma, R.N., Davis, DD and allagoa, M. PAH removal from Nigerian oil-based drill-cuttings with spent oyster mushroom (Pleurotus ostretus) substrate. Journal of Food, Agriculture and Environment 2010: 8 (3 &4), 914 -919.

36. Rojas-Avelizapa, N.G., Roldan-Carrillo, T., Zegarra-Martinez, H., Munoz-Colunga, A.M and Fernadez-Linares A field trail for an ex-situ bioremediation of a drilling mud-polluted site. Chemospher, 2007: 66, 1595 – 1600.

37. Martin, J.A., Moreno, J.L., Hernandez, T and Garcia, C

Bioremediation by composting of heavy oil refinery sludge in semiarid conditions. Biodegradation, 17:, 251 – 261.

38. Al-Nasrawi, H Biodegradation of Crude Oil by Fungi Isolated from Gulf of Mexico. J Bioremed Biodegrad 2012, 3:4

39. Mandal, A.K., Sarma, P. M., Singh, B., Jeyaseelan, C.P., Channasshettar, V.A., Lal, B and Datta, J bioremediation : an environment friendly sustainable biotechnological solution for remediation of petroleum hydrocarbon contaminated waste. ARPN Journal of Science and Technology, 2012: 2, 1-12

40. Stevenson, F.J Humus Chemistry, 2004. Wiley & Sons

41. Obayori, O.S., Ilori, M.O., Adebusoye, S.A., Amund, O.O and Oyetibo, G.O Microbial population changes in tropical agricultural soil experimentally contaminated with crude petroleum. African Journal of Biotechnology, 2008: 7 (24), 4512-4520.

42. Corwin, D.L and Lesch, S.M. Apparent soil electrical conductivity measurements in agriculture. Computers and Electronics in Agriculture, 2005: 46, 11–4

Chapter 2

Wellbore Stability in Oil and Gas Drilling with Chemical-Mechanical Coupling

Chuanliang Yan[1,2], Jingen Deng[1], and Baohua Yu[1]

[1]State Key Laboratory of Petroleum Resource and Prospecting, China University of Petroleum, Beijing 102249, China

[2]Department of Petroleum Engineering, China University of Petroleum, Beijing 102249, China

ABSTRACT

Wellbore instability in oil and gas drilling is resulted from both mechanical and chemical factors. Hydration is produced in shale formation owing to the influence of the chemical property of drilling fluid. A new experimental method to measure diffusion coefficient of shale hydration is given, and the calculation method of experimental results is introduced. The diffusion coefficient of shale hydration is measured with the downhole temperature and pressure condition, then the penetration migrate law of drilling fluid filtrate around the wellbore is calculated. Furthermore, the changing rules of shale mechanical properties affected by hydration and water absorption are studied through experiments. The relationships between shale mechanical parameters and the water content are established. The wellbore stability model chemical-mechanical coupling is obtained based on the experimental results. Under the action of drilling fluid, hydration makes the shale formation softened and produced the swelling strain after drilling. This will lead to the collapse pressure increases after drilling. The study results provide a reference for studying hydration collapse period of shale.

INTRODUCTION

Maintaining wellbore stability is an important issue in oil and gas industry [1–10]. In the process of drilling, the economic losses caused by wellbore instability reaches more than one billion dollar every year [11], and the lost time is accounting for over 40% of all drilling related nonproductive time [12]. It is also reported that shale account for 75% of all formations drilled by the oil and gas industry, and 90% of wellbore stability problems occur in shale formations [13–18]. When a well is drilled, the formation around the wellbore must sustain the load that was previously taken by the removed formation. As a result, an increase in stress around the wellbore and stress concentration will be produced [19–23]. If the strength of the formation is not strong enough the wellbore will be failure [24–28]. Wellbore stability is not only a pure rock mechanical problem, but also the interaction of drilling fluid and shale is a more important influence factor [29–35]. There are various chemicals in the drilling fluid which physically and chemically interact with shale formations. One hand, these interactions will result in the production of swelling stress [36–43]. On the other hand, it alleviates the mechanical strength of the wellbore wall rock [44–46]. Furthermore, it results in wellbore instability.

When studying the wellbore stability in shale, chemical factor must be combined with mechanical factor. Before the 1990s, the combinations are mainly on experimental study. Chenevert studied mechanical properties of shale after hydration since 1970s [44]. The results showed that the hydration would decrease the shale strength. After 1990s, the combinations came into a quantitative research stage. Yew et al. (1990) [29] and Huang et al. (1995) [47] combined shale hydrated effect quantitatively into the mechanical model based on thermoelasticity theory. Their method attributed the rock mechanical properties change with total water content. Take shale as a semipermeable membrane, Hale et al. (1993) [48, 49], Deng et al. (2003) [50], and Zhang et al. (2009) [51] introduced equivalent pore pressure to study interaction of shale and water base drilling fluid. Ghassemi et al. (2009) [52] proposed a linear chemo-thermo-poroelasticity coupling model, which considers the influence of chemical potential and temperature. Wang et al. (2012) [53, 54] built a fluid-solid-chemistry coupling model, in which they considered electrochemical potential, fluid flow caused by ion diffusion.

The chemical effect of drilling fluid on shale can be ultimately attributed to the variation of rock mechanical properties and stress around the wellbore. Water migration in shale is the basement of all wellbore stability models with chemical-mechanical coupling. A new experimental equipment to measure in situ water diffusion coefficient of shale is developed in this paper. And a sample model to evaluate time-dependent collapse pressure with chemical-mechanical coupling is presented.

EXPERIMENTAL RESEARCH ON THE HYDRATION OF SHALE

The free water and ion will penetrate into shale under the driving force of chemical potential and pressure difference between the pore fluid and drilling fluid [55–58]. Water content of shale changes by various mechanisms such as osmosis flow, viscous flow and capillary flow. Osmosis flow, driving force is due to chemicals and ions with different composition in drilling fluid and pore fluid. In order to evaluate the hydration of drilling fluid, the coefficient of water absorption and diffusion and the swelling ratio must be determined first [36].

Experimental Research on the Water Absorption of Shale

Experimental Equipment

Cherevent let one end face of shale sample contact with drilling fluid and the other end face wrapped up by plastic film, then he measured the water content increment in different location. But his experiment can only be conducted in room temperature and with zero confining pressure. But during drilling process in deep formation, it is in the condition of high temperature and high pressure. Shale hydration is influenced by temperature and pressure seriously, so his experimental result was inconsistent with actual drilling. In order to test the coefficient of water diffusion of shale, we developed an in situ test equipment of water diffusion coefficient which can fit the downhole temperature and pressure condition while drilling (Figure 1).

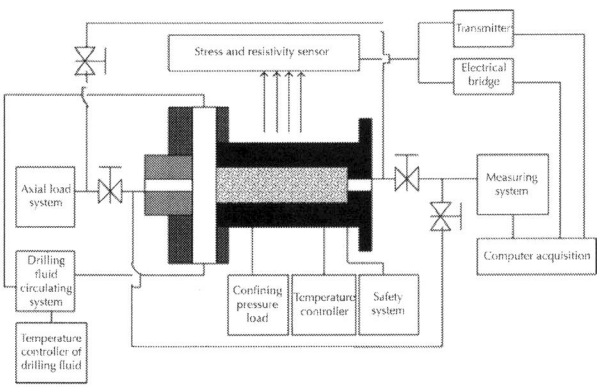

Figure 1: The experimental equipment sketch.

Technical parameters of this designed equipment are as the following.

- Temperature: room temperature to 150°C, which can imitate the temperature condition of the formation with 5000 meters depth.

- Pressure: confining pressure 0 MPa to 70 MPa, axial pressure 0 MPa to 200 MPa.

- Imitate the maximum differential pressure of drilling fluid with 10 MPa.

- Sample size: ⌀ 25 mm × 50 mm.

The experimental process are as follows.

- Determine the original water content of the rock samples first, wrap the samples with separation sleeve, and put into the core holder. Put the drilling fluid into the tank and check the test system to make sure it is in good condition.

- Turn on the temperature controller, warm the core samples to the same temperature with downhole condition. Then load the confining pressure and axial pressure to proper value and start timing.

- During the test, data acquisition control system is used to keep the test values constant.

- Cooling uninstall when the test time reaches the predetermined value (50 hours in this research), remove the rock samples quickly, and measure the water content at different distance from the end face.

Coefficient of Water Diffusion

According to conservation of mass, water diffusion equations can be established. Supposing q is mass flow rate of the water diffusion, $C_s(r, t)$ is the weight percentage of water at the time t and distance r away from the well axis; according to conservation of mass requirement, the following equation can be presented:

$$\nabla q = \frac{\partial C_S}{\partial t}.$$

(1)

Consider that

$$q = D_{eff} \nabla C_S,$$

(2)

Where ∇ the gradient operator and D_{eff} is the coefficient of water diffusion.

According to the above equations, water diffusion equation can be established as follows:

$$\frac{\partial C_S}{\partial t} - \frac{1}{r}\frac{\partial}{\partial r}\left(r\frac{\partial C_S}{\partial r}\right) D_{eff} = 0.$$

(3)

And the boundary conditions are

$$t = 0, \quad r_w \le r \le \infty, \quad C_S = C_0,$$

$$t > 0, \quad r = r_w, \quad C_S = C_{df},$$

$$t > 0, \quad r \longrightarrow \infty, \quad C_S = C_0,$$

(4)

Where r_w is the wellbore radius; C_{df} is the saturated water content of shale; C_0 is the original water content.

Sign $u = r / \sqrt{D_{eff}t}, C_s(r,t) = \phi(u)$, then the following equations can obtain that

$$\frac{\partial C_S}{\partial t} = \frac{d\phi}{du} \cdot \frac{\partial u}{\partial t} = \frac{d\phi}{du} \left(-\frac{1}{2}u\right) t^{-1},$$

$$\frac{\partial C_S}{\partial r} = \frac{d\phi}{du} \cdot \frac{\partial u}{\partial r} = \frac{d\phi}{du} \frac{1}{\sqrt{D_{eff}t}},$$

$$\frac{\partial^2 C_S}{\partial r^2} = \frac{d}{dr} \left(\frac{d\phi}{du} \frac{1}{\sqrt{D_{eff}t}}\right) = \frac{1}{D_{eff}t} \frac{\partial^2 \phi}{\partial u^2} \tag{5}$$

Insert (5) to (3), then (6) can be obtained:

$$\phi'' = -\left(\frac{u}{2} + \frac{1}{u}\right)\phi'. \tag{6}$$

Equations (7) and (8) can obtain by integrating (6) that

$$\phi' = -Au^{-1}e^{-u^2/4}, \tag{7}$$

$$C_S(r,t) = \phi(u) = \frac{A}{2} \int_{u^2/4}^{+\infty} x^{-1}e^{-x}dx + B. \tag{8}$$

Combining (8) and (4) the following equations can obtain that.

$$A = \frac{2\left(C_{df} - C_0\right)}{\int_{r_w^2/4D_{eff}t}^{+\infty} x^{-1}e^{-x}dx},$$

$$B = C_0. \tag{9}$$

Thus, the water content of shale formation around the wellbore can be written as follows:

$$C_S$$

$$= C_0 + \left(C_{df} - C_0\right)$$

$$\times \left[1 + \frac{2}{\pi} \int_0^\infty e^{-D_{eff}\zeta^2 \cdot t} \frac{J_0(\zeta r) N_0(\zeta r_w) - N_0(\zeta r) J_0(\zeta r_w)}{J_0^2(\zeta r_w) + N_0^2(\zeta r_w)} \right.$$

$$\left. \cdot \frac{d\zeta}{\zeta}\right], \tag{10}$$

Where J_0 () and N_0 () are the zero order of Bessel's functions of group one and two, respectively.

In a short period of time after drilling and within a short distance from the wellbore wall, (10) can be simplified to

$$C_S = C_0 + \left(C_{df} - C_0\right) \sqrt{\frac{r_w}{r}} \, \mathrm{erfc}\left(\frac{r - r_w}{2\sqrt{D_{\mathrm{eff}} t}}\right). \tag{11}$$

The water diffusion character of shale is measured using this designed experiment equipment. All the shale core samples used in this paper were collected from Bohai Bay Basin of China. The drilling fluid which contacted the shale in this experiment was KCL drilling fluid. The experimental confining pressure was 20 MPa, and the differential pressure of the fluid was 6 MPa. Core samples were taken out after 50 hours and then cut into pieces to measure the water content of each piece. Three samples were tested in this research. The experimental results of core sample 1-1 are shown in Figure 2. Substituting the experimental results into (11), the coefficient of water diffusion of shale can be obtained. All the calculated water diffusion coefficients and clay mineral contents of these three samples are shown in Table 1. Smectite is the mineral most prone to hydration [59, 60], and the water diffusion coefficient is higher with more smectite.

Table 1: Clay mineral contents and water diffusion coefficient of shale

Core no.	Clay mineral total contents (%)	Clay mineral relative contents (%)				D_{eff} (cm²/h)
		Smectite	Illite	Kaolinite	Chlo-rite	
1-1#	21.68	45	19	17	19	0.0238
1-2#	29.28	51	24	13	12	0.0247
1-3#	26.68	32	20	26	22	0.0184

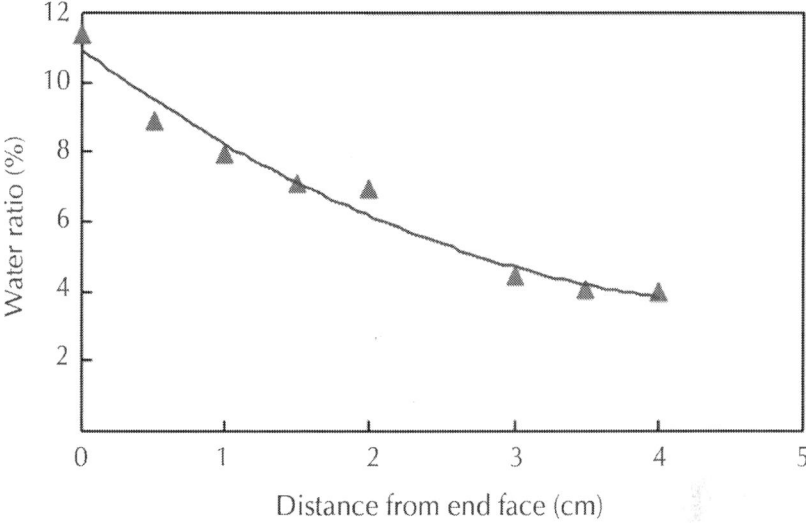

Figure 2: Experimental results of water diffusion in shale.

Chemical Effect of Drilling Fluid on Shale Mechanical Properties

The mechanical properties of shale can be altered seriously after contacting with drilling fluid. Existing forms of water in shale mainly include water vapor, solid water, bound water, adsorption water (film water), capillary water, and gravity water (free water) (Figure 3). Owing to the direct contact with drilling fluid around the wellbore, the free water of drilling fluid diffuses into shale under physical and chemical driving force. During the drilling process, the absorption water will increase, and the diffusion layer of rock particle will thicken, which will cause volume increase of shale and produce swelling stress. In order to calculate the swelling stress caused by hydration, the relation between water absorption and the swelling must be researched first through experiments. The experiment methods are similar to that used by Yew et al. [29]. The experimental results are shown in Figure 4.

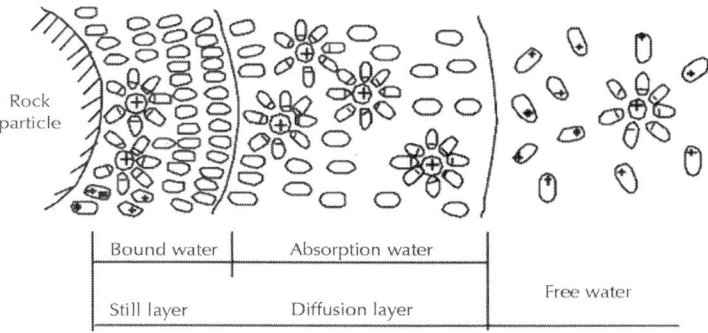

Figure 3: Water existing states in shale [63].

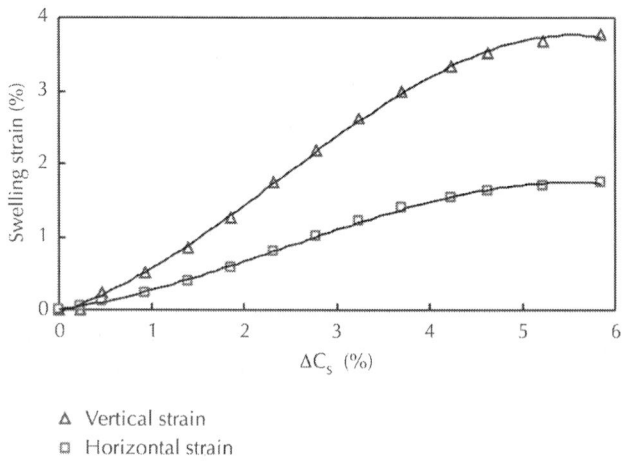

△ Vertical strain
□ Horizontal strain

Figure 4: Experimental results of shale swelling.

The experiment results show that the swelling strain in the direction perpendicular to the deposition surface is larger than that of the parallel direction, which is resulted from the difference of drainage and stress conditions in different directions in sedimentation [61, 62]. The relationship of water content and swelling strain is as follows:

$$\varepsilon_v = -324.8(\Delta C_S)^3 + 23.5(\Delta C_S)^2 + 0.37\Delta C_S,$$

$$\varepsilon_h = -150.8(\Delta C_S)^3 + 10.9(\Delta C_S)^2 + 0.17\Delta C_S, \qquad (12)$$

Where ε_v and ε_h are the swelling strain in the direction perpendicular and parallel to the deposition surface, respectively; ΔC_S is the water content increment.

When contacting with drilling fluid, water sensitivity minerals of shale will absorb water and make a chemical reaction. Clay mineral in shale will react with the ions in drilling fluid [63, 64]:

$$K_{0.9}Al_{2.9}Si_{3.1}O_{10}(OH)_2 + nH_2O$$

$$\longrightarrow K_{0.9}Al_{2.9}Si_{3.1}O_{10}(OH)_2 nH_2O. \qquad (13)$$

The properties of shale particle will change by above chemical reaction, and then weaken the cohesive force between shale particles, which will soften the shale and weaken the strength.

In order to evaluate wellbore stability after shale hydration, the mechanical properties of shale after hydration must be researched. In order to ensure the uniformity of the core samples used in the experiment, the compressive acoustic wave velocities of core samples were tested. Only the samples whose velocity is close are chosen. The core samples are shown in Figure 5. The samples are immersed in KCL drilling fluid at the temperature of 55°C. MTS-816 rock test system (Figure 6) was adopted to test shale mechanical properties. The experiment results are showed in Table 2.

Table 2: Experiment results of cores immersing in drilling fluid

Immersing time (h)	0	2	6	12	24	48	96	192

Water content increment ΔC_S (%)	0	0.82	2.03	2.70	3.28	4.46	5.11	5.68
UCS (MPa)	16.30	13.56	11.76	12.24	9.48	8.28	8.52	6.60
Poisson's ratio	0.26	0.30	0.32	0.30	0.34	0.38	0.41	0.40
Elastic modulus (MPa)	5518.37	4055.17	2881.29	2176.90	2111.45	1759.96	2165.25	1599.73

Figure 5: Standard samples of shale.

Figure 6: MTS-816 rock test system.

The relationship of rock mechanical parameters with water content increment is obtained by the experimental results:

$$E = 5.5 \times 10^3 e^{4.5\sqrt{\Delta C_S}},$$

$$v = 0.26 + 2.1\Delta C_S, \tag{14}$$

$$UCS = 16.3 - 1.49\Delta C_S, \tag{15}$$

Where E is the Young's modulus; v is the Poisson's ratio; UCS is the unconfined compressive strength.

Assume that the variation of shale cohesion with water content increment is the same with the variation of UCS:

$$\tau = \tau' - 1.49\Delta C_S, \tag{16}$$

Where τ is the cohesion; τ' is the initial cohesion before immersing, which is tested as 5.2 MPa.

The shear failure of wellbore obeys Mohr-Coulomb strength criterion. Mohr-Coulumb strength criterion can be expressed by principal stress [23]:

$$\sigma_1 = \sigma_3 ctg^2\left(45° - \frac{\varphi}{2}\right) + 2\tau \cdot ctg\left(45° - \frac{\varphi}{2}\right), \tag{17}$$

Where σ_1 and σ_3 are the maximum and minimum effective principal stresses, respectively; φ is the internal friction angle.

There is $\sigma_3 = 0$ in the uniaxial compression experiment. Inserting (15) and (16) into (17), the variation of internal friction angle with water content increment can be obtained:

$$\varphi = \arctan\left(\frac{16.3 - 1.49\Delta C_S}{10.4 - 2.98\Delta C_S}\right). \tag{18}$$

WELLBORE STABILITY MODEL WITH CHEMICAL-MECHANICAL COUPLING

Assuming that the shale is linear elastic material, the constitutive equation in plane strain is shown as follows:

$$\varepsilon_r = \frac{[\sigma_r - v(\sigma_\theta + \sigma_z)]}{E} + \varepsilon_h,$$

$$\varepsilon_\theta = \frac{[\sigma_\theta - v(\sigma_r + \sigma_z)]}{E} + \varepsilon_h,$$

$$\varepsilon_z = \frac{[\sigma_z - v(\sigma_r + \sigma_\theta)]}{E} + \varepsilon_v, \tag{19}$$

Where σ_r, σ_θ, and σ_z are radial, tangential, and vertical stresses, respectively; ε_r, ε_θ, and ε_z are radial, tangential, and vertical straines, respectively.

The formation is replaced by fluid column pressure after drilling. The original stress balance around the wellbore is broken. A new balance will be built. The stress balance equation is as follows [65]:

$$\frac{d\sigma_r}{dr} + \frac{\sigma_r - \sigma_\theta}{r} = 0, \tag{20}$$

Where σ_{ij} is the total stress tensor and f_i is the volume force.

The strain components and displacement components of the formation should meet the following geometric equation [65]:

$$\varepsilon_r = \frac{du}{dr},$$

$$\varepsilon_\theta = \frac{u}{r}, \tag{21}$$

Where ε_{ij} is total strain tensor and u_i is the displacement component.

Inserting (19) and (20) to (21), the following obtains that.

$$r\frac{d^2\sigma_r}{dr^2} + \left(3 - \frac{r}{E_1}\frac{dE_1}{dr} + \frac{2vr}{v^2 - 1}\frac{dv}{dr}\right)\frac{d\sigma_r}{dr}$$

$$+ \left(\frac{4v + 1}{v^2 - 1}\frac{dv}{dr} - \frac{1}{E_1}\frac{2v - 1}{v - 1}\frac{dE_1}{dr}\right)\sigma_r$$

$$= \frac{E_1(l + v)}{v^2 - 1}\frac{d\varepsilon_v}{dr} + \frac{E_1\varepsilon_v}{v^2 - 1}\frac{dv}{dr}. \tag{22}$$

The boundary conditions for drilling are as follows [66]:

$$\sigma_r = P_w, \quad r = r_w,$$
$$\sigma_r = S, \quad r = \infty, \tag{23}$$

Where P_w is under the fluid column pressure and S is the far field horizontal in situ stress.

Solving (22) and (23) using finite-difference method, stress distribution around the wellbore and its change rules with drilling time are obtained. Combined with Mohr-Coulumb failure criterion, the time-dependent collapse pressure can be obtained.

TIME-DEPENDENT COLLAPSE PRESSURE

Based on the above model and experimental results, the variations of mechanical parameters of shale around the wellbore and time-dependent collapse pressure are analyzed. The calculation parameters are as follows: well depth H=1800m, the initial water content C_0=4%, saturation water content C_{df}=11.4%, the water diffusion coefficient D_{eff}=0.0238cm²/h, and the wellbore radius r_w=10.8cm; the other parameters are obtained by the experimental results.

Figure 7 shows the variation of water content in shale formation around the wellbore with different open-hole time. The water content at the wellbore wall reaches to saturated state quickly after the wellbore is opened; in the same time, the water content would decrease with the increment of distance from the wellbore axis, and the decreasing rate is the highest near the wellbore wall. Thus, a hydrated area would develop around the wellbore. When the distance from the hole axis excesses 20 cm, the water content of shale almost no longer changes with the time increases, and the water content approaches the initial water content; in hydrated area, the longer the time, the more the shale water content when the distance is constant.

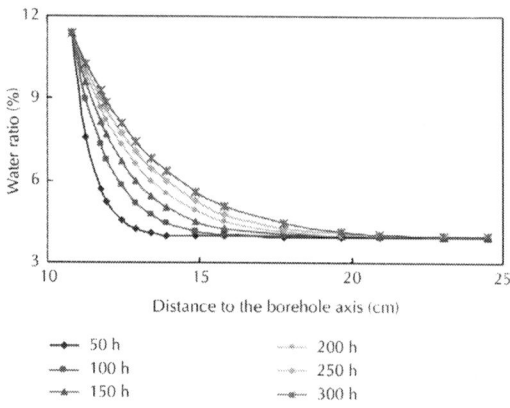

Figure 7: Water content distribution around the wellbore.

The distribution character of UCS of shale around the wellbore is presented in Figure 8. When a wellbore is opened, the UCS would decrease as the time increases. When the time is constant, the UCS would increase as the distance from the wellbore axis increases. The increasing rate near the wellbore wall is the highest.

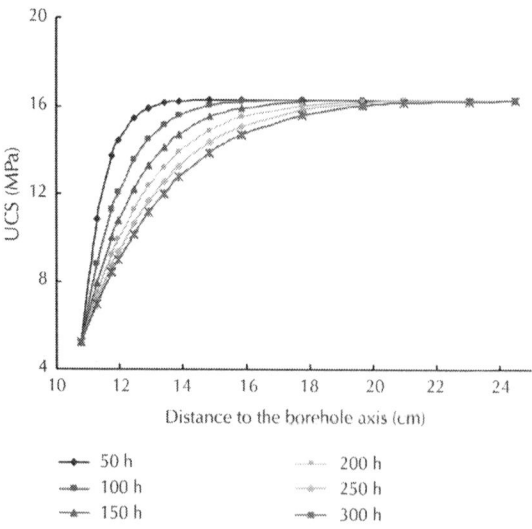

Figure 8: UCS distribution around the wellbore.

The variation of collapse pressure with drilling time under different water diffusion coefficients is shown in Figure 9. The results show that the collapse pressure increases rapidly in a short time after the wellbore opening due to shale hydration. Then the increasing rate of collapse pressure would decrease. At last, the collapse pressure would increase linearly with a very low increasing rate. After the wellbore is opened 10 days, the collapse pressure nearly no longer changes. According to Figure 9, after the wellbore opening, increasing the drilling fluid density to 1.45 g/cm^3 gradually is more beneficial for long-time wellbore stability when the water diffusion coefficient D_{eff}=0.0238cm^2/h. The higher the water diffusion coefficient is, the larger the increasing range of collapse is. On the other hand, the possibility of wellbore instability will be increasing with more smectite in shale.

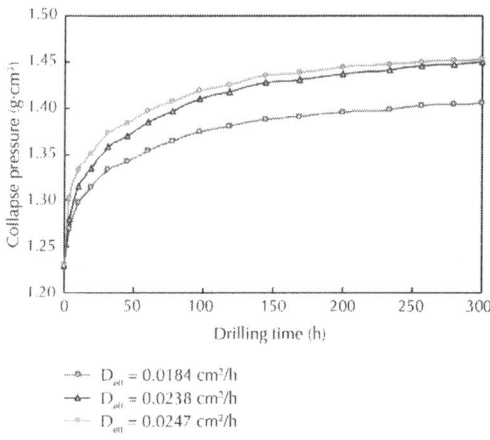

Figure 9: Time-dependent collapse pressure.

CONCLUSIONS

Water content at different distance from the end face of shale sample is measured using the designed equipment in the condition of downhole temperature and pressure.

The water content of shale at the wellbore wall reaches to saturated state quickly when the wellbore is opened; the water content of shale would decrease with the increment of distance from wellbore axis, and the decreasing rate is the highest near the wellbore wall.

Due to the impact of shale hydration, the strength of the circumferential formation around the well is gradually reduced with the increase of drilling time and increases with the increase of the distance away from the wellbore.

Collapse pressure of shale increases sharply in a short time after drilling and then slows down. The collapse pressure is basically steady after several days of the open-hole time. The initial stable wellbore may collapse with the increase of the open-hole time.

Shale containing more smectite is more prone to react with drilling fluid. The possibility of wellbore instability of shale is higher with more smectite as the increasing range of collapse pressure is larger.

ACKNOWLEDGMENTS

This work is financially supported by Science Fund for Creative Research Groups of the National Natural Science Foundation of China (Grant no. 51221003), National Natural Science Foundation Project of China (Grant no. 51134004 and Grant no. 51174219), and National Oil and Gas Major Project of China (Grant no. 2011ZX05009-005 and Grant no. 2011ZX05026-001-01).

REFERENCES

1.   W. B. Bradley, "Mathematical concept-stress cloud can predict borehole failure," Oil & Gas Journal, vol. 77, no. 8, pp. 92–101, 1979.

2.   F. J. Santarelli, E. T. Brown, and V. Maury, "Analysis of Borehole stresses using pressure-dependent, linear elasticity,"

International Journal of Rock Mechanics and Mining Sciences and, vol. 23, no. 6, pp. 445–449, 1986.

3.   B. S. Aadnoy, "Introduction to special issue on borehole stability," Journal of Petroleum Science and Engineering, vol. 38, no. 3-4, pp. 79–82, 2003. ·

4.   B. S. Aadnøy and M. Belayneh, "Elasto-plastic fracturing model for wellbore stability using non-penetrating fluids," Journal of Petroleum Science and Engineering, vol. 45, no. 3-4, pp. 179–192, 2004. ·

5.   L. Bailey, J. H. Denis, and G. C. Maitland, "Drilling fluids and wellbore stability current performance and future challenges," in Chemicals in the Oil Industry, P. H. Ogden, Ed., Royale Society of Chemistry, London, UK, 1991.

6.   J. S. Bell, "Practical methods for estimating in situ stresses for borehole stability applications in sedimentary basins," Journal of Petroleum Science and Engineering, vol. 38, no. 3-4, pp. 111–119, 2003. ·

7.   Y. Wang and M. B. Dusseault, "A coupled conductive-convective thermo-poroelastic solution and implications for wellbore stability," Journal of Petroleum Science and Engineering, vol. 38, no. 3-4, pp. 187–198, 2003.

8.   M. D. Zoback, C. A. Barton, M. Brudy et al., "Determination of stress orientation and magnitude in deep wells," International Journal of Rock Mechanics and Mining Sciences, vol. 40, no. 7-8, pp. 1049–1076, 2003.

9.   A. M. Al-Ajmi and R. W. Zimmerman, "Stability analysis of vertical boreholes using the Mogi-Coulomb failure criterion," International Journal of Rock Mechanics and Mining Sciences, vol. 43, no. 8, pp. 1200–1211, 2006.

10.   T. Al-Bazali, J. Zhang, M. E. Chenevert, and M. M. Sharma, "Factors controlling the compressive strength and acoustic properties of shales when interacting with water-based fluids," International Journal of Rock Mechanics and Mining Sciences, vol. 45, no. 5, pp. 729–738, 2008.

11.   M. E. Zeynali, "Mechanical and physico-chemical aspects

of wellbore stability during drilling operations," Journal of Petroleum Science and Engineering, vol. 82-83, pp. 120–124, 2012.

12. J. Zhang, J. Lang, and W. Standifird, "Stress, porosity, and failure-dependent compressional and shear velocity ratio and its application to wellbore stability," Journal of Petroleum Science and Engineering, vol. 69, no. 3-4, pp. 193–202, 2009.

13. F. K. Mody and A. H. Hale, "Borehole-stability model to couple the mechanics and chemistry of drilling-fluid/shale interactions," Journal of Petroleum Technology, vol. 45, no. 11, pp. 1093–1101, 1993.

14. G. Chen, M. E. Chenevert, M. M. Sharma, and M. Yu, "A study of wellbore stability in shales including poroelastic, chemical, and thermal effects," Journal of Petroleum Science and Engineering, vol. 38, no. 3-4, pp. 167–176, 2003.

15. L. C. Coelho, A. C. Soares, N. F. F. Ebecken, J. L. D. Alves, and L. Landau, "The impact of constitutive modeling of porous rocks on 2-D wellbore stability analysis," Journal of Petroleum Science and Engineering, vol. 46, no. 1-2, pp. 81–100, 2005.

16. H. C. H. Darley, "A laboratory investigation of borehole stability," Journal of Petroleum Technology, vol. 246, pp. 821–826, 1969.

17. R. T. Ewy and N. G. W. Cook, "Deformation and fracture around cylindrical openings in rock-I. Observations and analysis of deformations," International Journal of Rock Mechanics and Mining Sciences and, vol. 27, no. 5, pp. 387–407, 1990.

18. O. A. Helstrup, Z. Chen, and S. S. Rahman, "Time-dependent wellbore instability and ballooning in naturally fractured formations," Journal of Petroleum Science and Engineering, vol. 43, no. 1-2, pp. 113–128, 2004.

19. J. L. Yuan, J. G. Deng, Q. Tan, B. H. Yu, and X. C. Jin, "Borehole stability analysis of horizontal drilling in shale gas reservoirs," Rock Mechanics and Rock Engineering. In press.

20. J. C. Roegiers, "Well modeling: an overview," Oil and Gas Science and Technology, vol. 57, no. 5, pp. 569–577, 2002.

21. M. D. Zoback, D. Moos, L. Mastin, and R. N. Anderson, "Well bore breakouts and in situ stress," Journal of Geophysical Research, vol. 90, no. 7, pp. 5523–5530, 1985.

22. C. A. Barton, M. D. Zoback, and K. L. Burns, "In-situ stress orientation and magnitude at the Fenton Geothermal site, New Mexico, determined from wellbore breakouts," Geophysical Research Letters, vol. 15, no. 5, pp. 467–470, 1988.

23. E. Fjær, R. M. Holt, P. Horsrud, et al., Petroleum Related Rock Mechanics, Elsevier, 2nd edition, 2008.

24. R. Narayanasamy, D. Barr, and A. Milne, "Wellbore instability predictions within the cretaceous mudstones, clair field, west of shetlands," in Offshore Europe, Aberdeen, 2009.

25. E. T. Brown, J. W. Bray, and F. J. Santarelli, "Influence of stress-dependent elastic moduli on stresses and strains around axisymmetric boreholes," Rock Mechanics and Rock Engineering, vol. 22, no. 3, pp. 189–203, 1989.

26. P. A. Nawrocki, M. B. Dusseault, and R. K. Bratli, "Assessment of some semi-analytical models for non-linear modeling of borehole stresses," International Journal of Rock Mechanical & Mining Science, vol. 35, no. 4-5, pp. 522–531, 2002.

27. V. M. Maury and J. M. Sauzay, Borehole Instability: Case Histories, Rock Mechanics Approach, and Results, SPE, 1987.

28. J. C. Roegiers, "Well modeling: an overview," Oil and Gas Science and Technology, vol. 57, no. 5, pp. 569–577, 2002.

29. C. H. Yew, M. E. Chenevert, E. Martin, et al., "Wellbore stress distribution produced by moisture adsorption," SPE Drilling Engineering, vol. 5, no. 4, pp. 311–316, 1990.

30. E. van Oort, "On the physical and chemical stability of shales," Journal of Petroleum Science and Engineering, vol. 38, no. 3-4, pp. 213–235, 2003.

31. Z. Qiu, J. Xu, K. Lu, L. Yu, W. Huang, and Z. Wang, "Multivariate cooperation principle for well-bore stabilization," Acta

Petrolei Sinica, vol. 28, no. 2, pp. 117–119, 2007.

32. M. Yu, M. E. Chenevert, and M. M. Sharma, "Chemical-mechanical wellbore instability model for shales: accounting for solute diffusion," Journal of Petroleum Science and Engineering, vol. 38, no. 3-4, pp. 131–143, 2003.

33. A. Ghassemi and A. Diek, "Linear chemo-poroelasticity for swelling shales: theory and application,"Journal of Petroleum Science and Engineering, vol. 38, no. 3-4, pp. 199–212, 2003.

34. B. S. Aadnoy, "A complete elastic model for fluid-induced and in-situ generated stresses with the presence of a borehole," Energy Sources, vol. 9, no. 4, pp. 239–259, 1987.

35. L. Cui, Y. Abousleiman, A. H.-D. Cheng, and J.-C. Roegiers, "Time-dependent failure analysis of inclined boreholes in fluid-saturated formations," Journal of Energy Resources Technology, vol. 121, no. 1, pp. 31–39, 1999.

36. M. E. Chenevert and V. Pernot, "Control of shale swelling pressures using inhibitive water-base muds," in Proceedings of the 67th SPE Annual Technical Conference and Exhibition, pp. 27–30, SPE, New Orleans, La, USA, 1998.

37. P. T. Chee and G. R. Brian, Effects of Swelling and Hydration Stress in Shale on Wellbore Stability, SPE, 1997.

38. P. T. Chee and G. R. Brian, "Integrated rock mechanics and drilling fluid design approach to manage shale instability," in SPE/ISRM Rock Mechanics in Petroleum Engineering, 1998.
.

39. J. P. Simpson, H. L. Dearing, and D. P. Salisbury, "Downhole simulation cell shows unexpected effects on shale hydration on borehole wall," SPE Drilling Engineering, vol. 4, no. 1, pp. 24–30, 1989.

40. A. H. Hale, F. K. Mody, and D. P. Salisbury, "Experimental investigation of the influence of chemical potential on wellbore stability," in Drilling Conference, pp. 377–389, February 1992.

41. F. K. Mody and A. H. Hale, "Borehole stability model to

couple the mechanics and chemistry of drilling fluid shale interaction," in Proceedings of the SPE/IADC Drilling Conference, pp. 473–490, February 1993.

42. E. Oort, A. H. Hale, and F. K. Mody, "Manipulation of coupled osmotic flows for stabilisation of shales exposed to water-based drilling fluids," in Proceedings of the SPE Annual Technical Conference and Exhibition, pp. 497–509, October 1995.

43. C. P. Tan, M. Amanullah, F. K. Mody, and U. A. Tare, "Novel high membrane efficiency water-based drilling fluids for alleviating problems in troublesome shale formations," in Proceedings of the IADC/SPE Asia Pacific Drilling Technology, pp. 63–72, November 2002.

44. M. E. Chenevert, "Shale Alteration by Water Adsorption," Journal of Petroleum Technology, vol. 22, no. 9, pp. 1141–1148, 1970.

45. G. Z. Chen, A study of wellbore stability in shales including poroelastic, chemical, and thermal effects [Ph.D. dissertation], The University of Texas at Austin, 2001.

46. Y. H. Lu, M. Chen, Y. Jin, X. Q. Teng, W. Wu, and X. Q. Liu, "Experimental study of strength properties of deep mudstone under drilling fluid soaking," Chinese Journal of Rock Mechanics and Engineering, vol. 31, no. 7, pp. 1399–1405, 2012.

47. R. Z. Huang, M. Chen, and J. G. Deng, "Study on shale stability of wellbore by mechanics coupling with chemistry method," Drilling Fluid & Completion Fluid, vol. 12, no. 3, pp. 15–21, 1995.

48. A. H. Hale and F. K. Mody, "Borehole-stability model to couple the mechanics and chemistry of drilling-fluid/shale interactions," Journal of Petroleum Technology, vol. 45, no. 11, pp. 1093–1101, 1993.

49. A. H. Hale, F. K. Mody, and D. P. Salisbury, "Influence of chemical potential on wellbore stability," SPE Drilling and Completion, vol. 8, no. 3, pp. 207–216, 1993.

50. J. Deng, D. Guo, J. Zhou, and S. Liu, "Mechanics-chemistry coupling calculation model of borehole stress in shale formation and its numerical solving method," Chinese Journal of Rock Mechanics and Engineering, vol. 22, no. 1, pp. 2250–2253, 2003.

51. L. W. Zhang, D. H. Qiu, and Y. F. Cheng, "Research on the wellbore stability model coupled mechanics and chemistry," Journal of Shandong University, Engineering Science, vol. 39, no. 3, pp. 111–114, 2009.

52. A. Ghassemi, Q. Tao, and A. Diek, "Influence of coupled chemo-poro-thermoelastic processes on pore pressure and stress distributions around a wellbore in swelling shale," Journal of Petroleum Science and Engineering, vol. 67, no. 1-2, pp. 57–64, 2009.

53. Q. Wang, Y. Zhou, Y. Tang, and Z. Jiang, "Analysis of effect factor in shale wellbore stability," Chinese Journal of Rock Mechanics and Engineering, vol. 31, no. 1, pp. 171–179, 2012. ·

54. Q. Wang, Y. C. Zhou, G. Wang, H. W. Jiang, and Y. S. Liu, "A fluid-solid-chemistry coupling model for shale wellbore stability," Petroleum Exploration and Development, vol. 39, no. 4, pp. 475–480, 2012.

55. M. E. Chenevert and A. K. Sharma, "Permeability and effective pore pressure of shales," SPE Drilling & Completion, vol. 8, no. 1, pp. 28–34, 1993.

56. V. X. Nguyen, Y. N. Abousleiman, and S. K. Hoang, "Analyses of wellbore instability in drilling through chemically active fractured-rock formations," SPE Journal, vol. 14, no. 2, pp. 283–301, 2009.

57. E. van Oort, "Physico-chemical stabilization of shales," in Proceedings of the SPE International Symposium on Oilfield Chemistry, pp. 523–538, February 1997.

58. E. van Oort, "On the physical and chemical stability of shales," Journal of Petroleum Science and Engineering, vol. 38, no. 3-4, pp. 213–235, 2003.

59. E. Oort, A. H. Hale, F. K. Mody, and S. Roy, "Critical parameters in modelling the chemical aspects of borehole stability in shales and in designing improved water-based shale drilling fluids," in Proceedings of the SPE Annual Technical Conference & Exhibition, pp. 171–186, September 1994.

60. M. Chen, Y. Jin, and G. Q. Zhang, Petroleum Engineering Related Rock Mechanics, Science Press, Beijing, China, 2008.

61. C. E. Weaver, Clays, Muds, and Shales, Elsevier, 1989.

62. S. H. Ong, Borehole Stability [Ph.D. dissertation], The U. Of Oklahoma, 1994.

63. X. J. Zhu, "Water-weakening properties of soften rocks," Technology of Mineral Science, vol. 3-4, pp. 46–50, 1996.

64. C.-H. Yang, H.-J. Mao, X.-C. Wang, X.-H. Li, and J.-W. Chen, "Study on variation of microstructure and mechanical properties of water-weakening slates," Rock and Soil Mechanics, vol. 27, no. 12, pp. 2090–2098, 2006.

65. Z. L. Xu, Elastic Mechanics, Higher Education Press, Beijing, China, 1998.

66. J. G. Deng, "Calculation method of mud density to control borehole closure rate," Chinese Journal of Rock Mechanics and Engineering, vol. 16, no. 6, pp. 522–528, 1997.

Chapter 3

Evaluation of Control Methods for Drilling Operations with Unexpected Gas Influx

Liv A. Carlsen[a, c], Gerhard Nygaard[b, c], and Michael Nikolaou[d]

[a]IRIS, 4068 Stavanger, Norway

[b]IRIS, Thormøhlensgate 55, 5008 Bergen, Norway

[c]Department of Petroleum Engineering, University of Stavanger, Norway

[d]Department of Chemical and Biomolecular Engineering, University of Houston, Houston, TX, United States

ABSTRACT

This paper presents an evaluation of various control methods to be used during drilling operations where an unexpected gas influx occurs. In the event of an unexpected gas influx the current industry procedure is to control the pressure in the well manually. The drilling industry term for this manual procedure is well control. The focus of the paper is threefold. Firstly, to design an automatic sequence which is similar to the existing manual procedure. Secondly, to evaluate three different control algorithms for pressure control during an unexpected gas influx, and thirdly, to evaluate control parameter tuning needed when implementing different control algorithms.

The control methods have been evaluated on various drilling scenarios with unexpected gas influx, referred to as a kick. After a kick of reservoir gas has entered the well, automatic control of the well control choke and rig pump is applied to compensate for pressure fluctuations while circulating out the gas. A PI controller is designed to stabilize the well pressure by controlling the well control choke, an internal model controller (IMC) controls the pressure by manipulating the choke and the rig pump flowrate, and a model predictive controller (MPC) uses coordinated control of the choke and the pump flowrate to stabilize the well pressure. The model based controllers use a simple first order model of the well. Simulations are performed using a detailed flow model of the well to test the controller performance and robustness. Several cases with different amounts of gas influx are investigated.

The simulations show that it is feasible to control the pressure using automatic control of the choke valve and pump during an unexpected gas influx by use of all the presented control methods. The control methods are robust against changes in process conditions and disturbances, as they are able to handle several pressure levels and gas volumes without requiring re-tuning. However, since the pressure dynamics in the well are influenced when gas is entering the well, the model based controllers could probably be further improved if the models were updated after the gas influx occurred.

The results indicate that adaption of the automatic sequence to the current manual procedure is applicable. However, to avoid a reduction in downhole pressure when stopping the pump and shutting in the well, the automatic sequence may be further improved beyond what is feasible with manual operation.

INTRODUCTION

While drilling in the reservoir section of a well, pressure control is crucial. The well pressure in this section should stay above the pore pressure of the formation. If the pressure falls below the pore pressure, reservoir fluids or gas will leak into the well. This is called an influx, and if the influx is above a certain size it is termed a kick. If not handled properly, this may lead to a blowout situation, where the reservoir fluids flow uncontrolled into the wellbore and to the surface. On the other hand, if the pressure in the well is too high, this may cause loss of mud to the formation and well fracture.

A kick can be caused by too low mud density, such that the hydrostatic pressure in the well is below the pore pressure, or by transient effects like drillstring movement. There may be several indicators that a kick has occurred. One indicator is increased volume in the pit tank, which is the tank containing the drilling mud. If gas starts to leak into the well, the volume of the gas will occupy the annulus volume and push the drilling fluid upwards. The flow rate out of the annulus will then be higher than the flow rate pumped into the drillstring from the pit, causing the volume in the pit tank to increase. A quicker indicator is comparing the flow rate into and out of the well directly, not through the pit volume. This however requires a flow meter at the outlet of the well, which is usually not available. Other kick indications are sudden increase in drilling rate, change in pump pressure and reduction in drillpipe weight [1].

If the kick is above a certain size, the well needs to be shut in to stop the kick. The kick must then be circulated out in a controlled manner. This procedure, called well control, is a manual operation including sensor recordings, calculations and control performed by several members of the drilling crew. It involves control of the blow-out preventer (BOP), the rig pump and the well control choke, all located at different locations at the rig. The well control choke is adjusted manually in order to maintain a certain pressure in the well. This may be a difficult task due to large time-delays in the drilling process and the complex behavior of the multiphase flow. Automation of the well control procedure including pressure control with automatic choke adjustments would improve reliability and reduce the possibility of human errors.

The lack of automation level within the industry may have several reasons: firstly, the drilling operation is performed partly by the oil company, the drilling contractor and the service company. This means that the overall process overview is not fully understood by each of the companies involved in the drilling operation. This applies specifically to the service companies since they typically deliver equipment and personnel trained to operate their equipment. Because automation would mean less personnel involved, this would lead to reduced revenue for the service companies.

In other industries automation has evolved due to economic reasons, and since oil drilling has not reached sufficiently hard cost constraints no motivation for automation has been present. However, since automation also gives improved Health, Safety and Environment (HSE), new focus on automation of the drilling process can currently be seen.

In recent years, several drilling operations have been performed successfully using managed pressure drilling (MPD) techniques. Managed pressure drilling is a collective term for various techniques employed to manage the annular hydraulic pressure profile of the exposed wellbore [2]. A typical layout of an annulus backpressure type MPD system is described in [3] and [4], and control requirements for MPD systems are discussed in [5]. In [6] and [5] PID control methods are applied to control the pressure during MPD operations. Nonlinear model based control solutions and observers are addressed in [7] together with experimental results. The use of model predictive control is proposed as a multivariable control framework for dual-gradient drilling by [8] and for an annulus backpressure systems by [9]. A multilevel control approach of an MPD operation is presented in [10] and [11].

With managed pressure drilling systems, a new possibility of handling small kicks without shutting in the well and increasing the mud weight to increase the hydrostatic pressure is introduced. Instead, the influx can be stopped by increasing the well pressure through adjustment of the MPD choke. Several papers describe the need for new well control procedures during drilling operations utilizing equipment designed for managed pressure drilling including both annulus backpressure systems and dual gradient systems [7]. An evaluation of different approaches of initial responses to kick during MPD operations is given by [12]. For annulus backpressure systems, a well control method described as micro-flux control is described by [13] where the difference between flow in and flow out of the well is compared, and instant choke adjustments are made to maintain control over the well. An evaluation of well control procedures for dual gradient systems including riser-gas lift is given by [14]. In [15], a kick detection and attenuation algorithm is developed for a dual gradient system. Managed pressure drilling also introduces new techniques for kick detection and pore pressure estimation as described by [16] and [17].

The new techniques for kick handling during MPD operations are only applicable for relatively small kicks. When experiencing large kicks, it is still recommended to shut in the well and circulate out the kick using standard well control procedures and equipment. Only a very low percentage of wells today are drilled using MPD techniques. In most kick situations MPD equipment is not available, and standard well control procedures using the well control equipment must be applied.

In this paper, ideas and pressure control methods from MPD operations are applied to automatically control the pressure while performing standard well control procedures with the well control equipment. It is demonstrated that simple control methods can be applied for the purpose. In the event of a larger kick event the standard industry procedure today is still to control the pressure in the well manually. The focus of the paper is threefold. Firstly, to design an automatic sequence which is similar to the existing manual procedure. Secondly, to evaluate three different control algorithms for pressure control during an unexpected gas influx, and thirdly, to evaluate control parameter tuning needed when implementing different control algorithms.

Section 2 of this paper gives an overview of a conventional drilling system and describes how standard well control procedures are performed manually today. In Section 3, the details for the well test case used during the simulations are given. A first-order approximation model of the system to be used in the controller design is presented. Section 4 presents three different control methods for controlling the pressure during well control incidents, a PI control design, an internal model control algorithm and a model predictive controller. The controllers are evaluated by running different simulation scenarios on a high-fidelity simulator in Section 5. The next section discusses the tuning of parameters of the various controllers, and conclusions and ideas for further work are given in Section 7.

DRILLING AND WELL CONTROL

A schematic of a conventional drilling system is shown in Fig. 1. The system consists of a rotating drillstring with a drilling bit at the bottom. The volume that develops around the drillstring as the well is drilled is referred to as the annulus. Drilling mud is pumped from a pit tank, down the drillstring, through the bit, up the annulus, and back into the pit tank. The purpose of the drilling mud is to transport cuttings from the bottom of the well, and to create a certain pressure balance in the well. The pressure in the well during a conventional drilling operation is managed by the density and the flow rate of the drilling fluid.

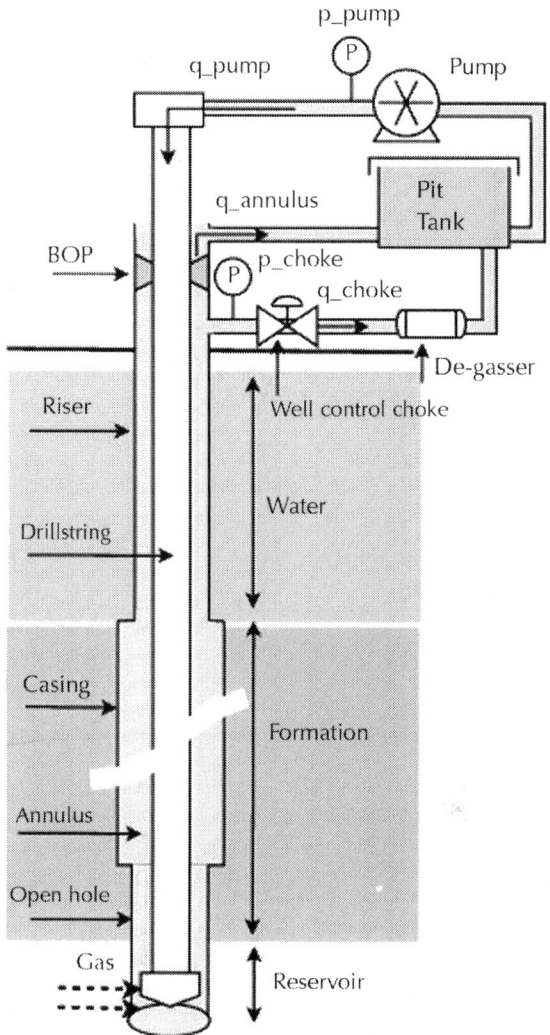

Figure 1: Drilling system.

The well control equipment is depicted in Fig. 1. A blow-out preventer is located on the top of the annulus, or at the sea floor on floating rigs. The purpose of the BOP is to seal the annulus in a kick situation, to prevent reservoir fluids from flowing uncontrolled out of the well. When the BOP is closed, the circulation path is directed through the choke line, located below the BOP, though the well control choke, through the de-gasser, and into the pit tank.

During a conventional drilling operation, the only secure method to stop a kick is to shut in the well. A typical shut-in procedure [1] is to stop the pumps, open the well control choke, close the blow-out preventer, and close the well control choke. This is termed a "soft shut-in". The alternative is to close the blow-out preventer on a closed choke, which is known as a "hard shut-in".

When the well has been shut in, the pressure in the well will start to increase. This is because as long as the well pressure at the location of the influx is below the pore pressure, gas will continue to flow into the well, filling the constant volume in the closed annulus, which will cause the pressure increase. When the pressure at the influx location has increased above the pore pressure, the influx stops and the pressure stabilizes.

Once the influx is stopped and the pressure is stable, the procedure continues by circulating out the influx that has entered the annulus. The influx is circulated out through the choke line below the blow-out preventer. While circulating out a gas kick, the gas must be allowed to expand to avoid rupturing the wellbore. At the same time, the pressure at the bottom of the hole must be kept above the pore pressure to prevent additional influx from the reservoir. The goal of classic well control procedures is to maintain a constant downhole pressure while circulating out the kick. The pressure is controlled by adjusting the well control choke at the outlet. However, measurements of the downhole pressure are usually not available. Only topside measurements such as pump pressure, also referred to as the drillpipe or standpipe pressure, and measurements of the pressure at the inlet of the well control choke, termed choke or casing pressure, can be used.

Classic well control procedures apply these pressure measurements to indirectly control the downhole pressure. These procedures are based on the assumption that when using slow pump rates, the frictional pressure drop in the annulus can be neglected, focusing on frictional pressure loss in the drillpipe only.

When the pump is running at constant rate, the pressure at downhole will remain constant if the pump pressure is constant. But when ramping up or down the pump, keeping the pump pressure constant will not give a constant downhole pressure due to frictional pressure changes in the drillpipe.

During pump ramp up and ramp down the downhole pressure will remain approximately constant if the choke pressure is kept constant. This is however not the case in wells with a large frictional pressure drop in the annulus, for example in slim holes. When starting up the pump in wells with large frictional pressure drop in the annulus, keeping the casing pressure constant will cause an increase in the downhole pressure. But if the pressure increase is not too high, it will simply be a safety barrier towards the pore pressure.

The most applied well control procedures are the Driller's Method and the Wait and Weight Method [1]. The Driller's Method is simple and requires minimal calculations. The procedure, depicted in Fig. 2, starts by bringing the pump to a desired circulation rate, keeping the casing pressure constant at the shut-in casing pressure, which is the casing pressure when the pressure has stabilized after a shut-in. When the pumps are running at the desired rate, the influx is displaced while keeping the drillpipe pressure constant. In order for the well to regain overbalance, the mud weight must be increased. This is referred to as to kill the well. The well is in overbalance when the hydrostatic pressure caused by the drilling fluid in the wellbore is higher than the pore pressure. The Driller's Method continues by circulating the heavier kill-weight mud through the drillpipe and finally displacing the heavier mud to the surface. The well should then be in overbalance, and it is safe to open the blow-out preventer and continue the drilling operation.

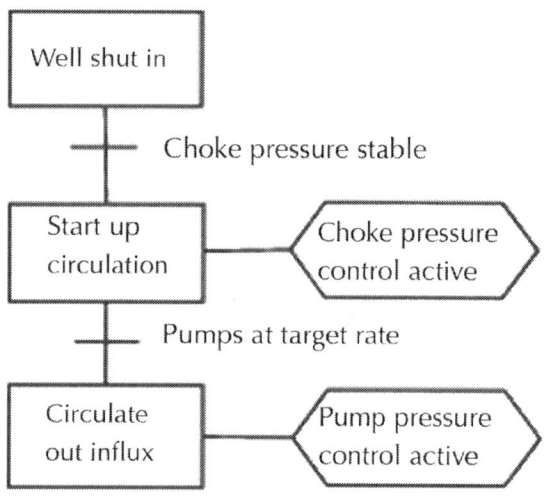

Figure 2: Well control.

Instrumentation

Instrumentation during drilling operations is poor. The pressure at the bit is sometimes measured, but the signal is sent with mud pulses to the surface, which limits bandwidth significantly. The signal will thus have a delay of up to several seconds and cannot be utilized for control purposes. The pump pressure p_{pump} is measured with a pressure sensor placed along the path between the mud pump and the drillpipe, as depicted in Fig. 1. In a well control situation, when the circulation path goes through the well control choke, the choke pressure p_{choke} is measured with a pressure sensor placed before the choke, also depicted in Fig. 1. The pump flow rate, q_{pump}, is measured by counting the pump strokes per minute. When the volume of each stroke is known, the pump strokes per minute can be converted to volume flow rate per minute. Unknown pump efficiency and pump leakage can cause incorrect flow rate readings. The flow rate out of the well, $q_{annulus}$, is measured using a flow paddle placed in the return channel from the annulus to the pit tank. This is an inaccurate flow measurement.

CASE DESCRIPTION

Simulation Model

The simulations performed in this paper are evaluated using a detailed multiphase fluid model [18], modeling both the wellbore dynamic pressure and velocity effects, in addition to the transient behavior of the influx from the reservoir. The model assumes dispersed bubble flow and calculations are performed using a drift-flux model. Temperature effects are included in the model. However, the temperature interaction between the well fluids and formation fluids are not handled. Reference to the model is given in [9]. The simulation model has been verified by several real-time applications [19] and [20].

An inclined well of length 2300 m is used as a test case, with a water based drilling fluid with density 1475 kg/m³. Further details of the well, the drilling fluid and the reservoir conditions are given in Table 1.

Table 1: Well and reservoir data

Parameter	Value
Well total length	2300 m
Well vertical depth	1720 m
Reservoir height	2 m
Reservoir permeability	100 mD
Reservoir pore pressure	262 bar
Reservoir porosity	0.18
Skin factor	0.13
Fluid density	1.475 s.g.

Control Objectives and Variables

The control objective is to keep the pressure at the bottom of the well constant during a well control situation. When measurements of the downhole pressure are available, the downhole pressure can be controlled directly. When these measurements are not available, the control objective is to keep the choke pressure constant while starting up the pump, and keeping the pump pressure constant while circulating out the gas. The main manipulated variable is the choke opening of the well control choke, situated at the outlet of the well as depicted in Fig. 1. The internal model control (IMC) method and the model predictive control (MPC) method used in this paper have the pump flow rate as a second manipulated variable. For the PI controller, the pump flow rate will be a disturbance. Other unmeasured disturbances are gas flow and expansion in the annulus and through the choke, which will cause pressure fluctuations. The control constraints are the pore pressure (lower constraint) and the fracture pressure (higher constraint). The margins towards the constraints should be high, since the pore- and fracture pressure are uncertain. In this paper, the control objective is to stay within ±2 bar of the pressure set point.

First-order Model Approximation

The detailed model used for the simulations has been used in a nonlinear MPC in [9]. However, these detailed models need several parameters that are typically unknown in a real drilling operation.

A low-order model [21] fit for control purposes has been developed and used in control applications for managed pressure drilling [6] and [7], and observers for downhole pressure [22]. However, the model has currently only been used in laboratory facility setups. In addition, the model requires certain parameters such as the drillstring frictional pressure drop, and at the current drilling rigs this parameter cannot easily be measured. The model is a one-phase model and does not include a model for gas.

For an industrial application the search for a sufficiently accurate control design is needed. Even the simplest control design based on a first order model has not been evaluated sufficiently. Experiences from other industries using MPC approaches indicate that a first order model is often sufficient. In this paper, first-order step response models are therefore used in the controller design, similar to what is applied in [5].

The pressure step response when exposed to a step change in the choke opening or pump flow rate can be approximated by

$$p(s) = \frac{K_c}{\tau_c s + 1} z_c(s) + \frac{K_p}{\tau_p s + 1} q_p(s)$$

(1)

Where p is the pressure, z_c is the choke opening, q_p is the pump rate, K_c and K_p are the process gains, and τ_c and τ_p are the time constants. The pressure p can be one of each pressure: choke pressure, pump pressure or downhole pressure. The process is non-linear, since the process gains K_c and K_p, and time constants τ_c and τ_p are dependent on the choke opening and flow rate. The parameters increase with increasing flow rate and decreasing choke opening.

The simulation model is here used as reality in order to define the parameters in the first order models. The experiments to define the parameters of the first order model can also be performed in a real drilling operation.

Fig. 3 shows the pump pressure, choke pressure and downhole pressure step responses to various step changes in the choke opening, when the pump rate is kept constant at 1000 l/min. Fig. 4 shows the pressure response with step changes in the pump flow rate. In these cases, the choke opening is constant at 10%. An example of first order model approximations to the pressure step responses after a step change in the choke opening is shown in Fig. 5. Fig. 6 shows first order model approximations to the pressure step responses after a step change in the pump flowrate.

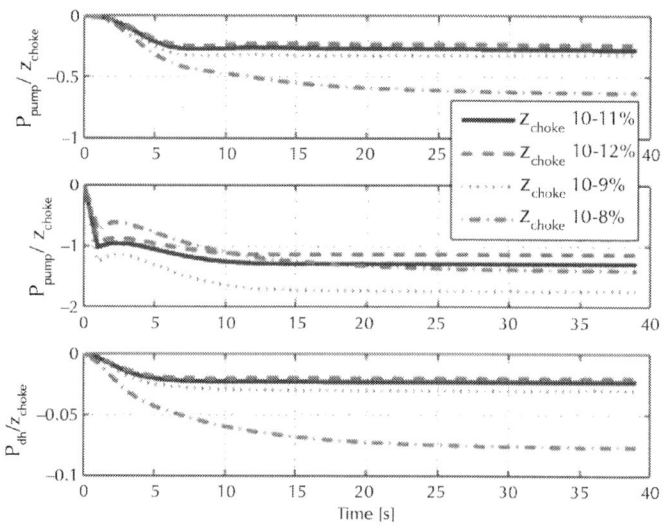

Figure 3: Pressure step responses to various step changes in choke opening. Constant flow rate 1000 l/min.

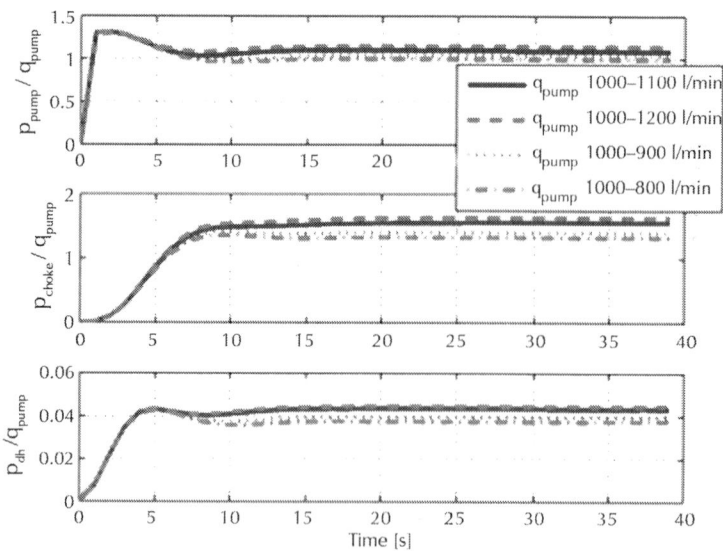

Figure 4: Pressure step responses to various step changes in pump flow rate. Constant choke opening 10%.

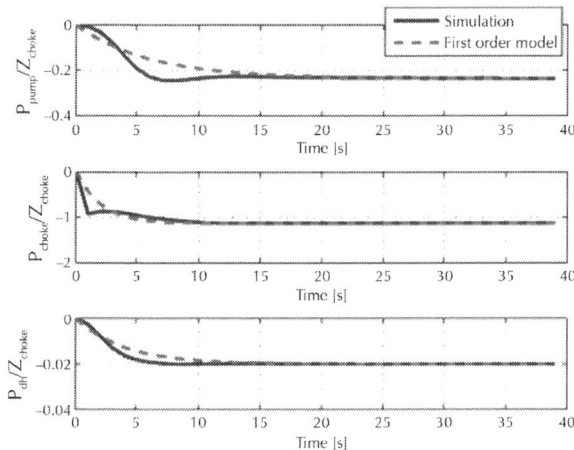

Figure 5: Pressure step response to step change in choke opening from 10% to 12%. Blue solid line is from simulations, and red dashed line is a first order model approximation. (For interpretation of the references to color in this figure legend, the reader is referred to the web version of the article.).

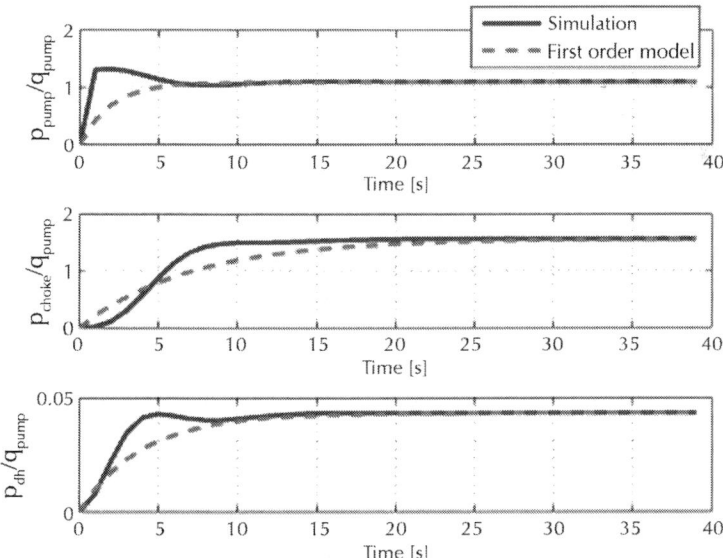

Figure 6: Pressure step response to step change in flow rate from 1000 to 1100 l/min. Blue solid line is from simulations, and red dashed line is a

first order model approximation. (For interpretation of the references to color in this figure legend, the reader is referred to the web version of the article.).

CONTROL SCHEMES SETUP

Three different control methods have been evaluated for pressure control during a well control situation. The first is a standard PI control algorithm that controls the pressure by manipulating the choke opening. The second method is an IMC algorithm [23] controlling the pressure by manipulating the choke opening and the pump flowrate, using the first-order model of the well system from Eq. (1). The IMC structure controls pressure by adjusting the choke using a PI controller. Pump flow rate is controlled close to its setpoint using an integral-only controller. The entire control scheme is essentially a feedback/feedforward loop, using the choke as feedback manipulation and the pump flow rate as measured disturbance for feedforward corrections. The third control method is a linear MPC method [24]. This controller applies coordinated control of both the choke opening and the pump rate to control the pressure. The first-order model from Eq. (1) is used in the plant model. All three controllers have been developed for controlling choke pressure, pump pressure or downhole pressure. A list of the controllers and control variables is given in Table 2.

Table 2: Controllers

Controller type	Manipulated variables	Controlled output
PI	Choke opening	Choke pressure, pump pressure or downhole pressure
IMC	Choke opening and pump rate	Choke pressure, pump pressure or downhole pressure
MPC	Choke opening and pump rate	Choke pressure, pump pressure or downhole pressure

The three main parts of the well control sequence are well shut-in, pump start-up and circulation. In each of these parts the controlled output is different. In the well shut-in sequence the choke valve and main pump are operated using only open loop control. The main pump is ramped down at a pre-defined schedule and the valve is as a main rule closed when the pump is fully stopped. In the pump start-up sequence the controlled output is the choke pressure, and during the circulation sequence the controlled output is the pump pressure. For the IMC and the MPC controller the pump flowrate is a manipulated variable in addition to the choke valve opening. If the downhole pressure is available during the well control sequence then the controlled output is downhole pressure. An overview of the various sequences and the controlled output for each sequence is given in Table 3. Table 4 shows the form of the controllers, and the control values for each controller type and controlled output are given in Table 5. The tuning of the controllers is based on simulated response tests for each controller structure. A further discussion on the controller tuning is presented in Section 6.

Table 3: Well control sequences

Part of sequence	Controlled output without downhole measurement	Controlled output with downhole measurement
Shut in	Open loop	Open loop
Circulation start-up	Choke pressure	Downhole pressure
Circulation	Pump pressure	Downhole pressure

Table 4: Controller form

Controller type	Controller form
PI	$C(s) = K_p\left(1 + \dfrac{1}{\tau_I s}\right)$
IMC	$C(s) = \begin{bmatrix} \dfrac{\tau_c}{K_c\lambda_1}\left(1 + \dfrac{1}{\tau_c s}\right) & -\dfrac{\tau_c}{K_c\lambda_2}\left(1 + \dfrac{1}{\tau_c s}\right)\dfrac{K_p}{\tau_p s + 1} \\[2ex] \dfrac{1}{\lambda_2 s} & 0 \end{bmatrix}$
MPC	$\displaystyle\min_{\Delta u(k),\dots\Delta u(m-1+k)}\left\{\sum_{j=1}^{p}\left| [w\left(y(k+i)-y^{sp}\right)]\right|^2 + [r_1\Delta u_1(k+i)^2 + r_2\Delta u_2(k+i)^2 + [v_1(u_1(k+i)-u_{1target}(k+i))^2 + v_2(u_2(k+i)-u_{2target}(k+i))^2]\right\}$
	$\text{s.t. } y_{min} \le y(k+i) \le y_{max}$

Table 5: Controller parameters

Controller type	Controlled output		
	Choke pressure	Pump pressure	Downhole pressure
PI	$K_p = -0.068$	$K_p = -2.3$	$K_p = -25$
	$\tau_I = 0.4$	$\tau_I = 10$	$\tau_I = 8$
IMC	$\lambda_1 = 0.86$	$\lambda_1 = 0.83$	$\lambda_1 = 0.7$
	$\lambda_2 = 0.95$	$\lambda_2 = 0.95$	$\lambda_2 = 0.9$
MPC	p = 100	p = 100	p = 100
	m = 2	m = 2	m = 2
	w=0.85	w=0.3	w=0.9
	$r_1 = 0.1$	$r_1 = 0.1$	$r_1 = 0.1$
	$r_2 = 0.1$	$r_2 = 0.1$	$r_2 = 0.1$
	$v_1=0.0$	$v_1=0.0$	$v_1=0.0$
	$v_2=0.001$	$v_2=0.0015$	$v_2=0.003$

CONTROL SCHEMES EVALUATION

The three controllers presented in the previous section have been evaluated for automatic pressure control during a well control procedure by running simulations using the detailed fluid model and the well test case described in Section 3. The lower pressure constraint is the pore pressure, and the upper constraint is fracture pressure at the casing shoe. However, the pore and fracture pressure in real operations are not known exactly, so the controller should always stay as close to the reference value as possible. The control objective is therefore to keep the pressure within ±2 bar of the reference pressure during the procedure.

In Simulation 1, a kick incident occurs during a conventional drilling operation with no downhole pressure measurements available. The well is shut in and the kick is circulated out with automatic control of the choke and pump pressure. The simulation setup in Simulation 2 and Simulation 3 is identical to the first, except the pore pressure and permeability of the reservoir is increased respectively, in order to test the controller robustness. In Simulation

4, measurements of the downhole pressure are available, and the kick is circulated out while controlling the downhole pressure explicitly.

Simulation 1: Well Control using Topside Pressure Measurements

This section presents results from simulations of a well control procedure where the downhole pressure measurements are not available, which is the case in most drilling operations. The procedure uses the choke and pump pressure measurements during circulation start-up and while circulating out the kick, as explained for the standard well control procedures in Section 2, only with automatic instead of manual control of the pressure.

Initially, a conventional drilling operation in the reservoir zone is going on. The pore pressure at the bottom of the hole is 262 bar. The blow-out preventer is open and the well control choke is closed. The pump is running at 2000 l/min, pumping drilling mud down the drill string, up the annulus and out through the open top of the annulus into atmospheric pressure. This pump rate gives a downhole pressure of 258 bar which is below the critical pore pressure at 262 bar, resulting in a situation where reservoir gas starts to flow into the bottom of the well. When the gas flows into the annulus, filling the fixed volume, the mud flow rate out of the annulus is higher than the mud flow rate pumped into the drillstring, and the pit tank volume will then start to increase. When the pit gain has increased above 2 m^3, this is defined as a kick situation, and the well control procedure with automatic pressure control is initiated.

The choke pressure, pump pressure and downhole pressure during the whole simulation is shown in Fig. 7, and the time period for various sequences in the simulation are listed in Table 6.

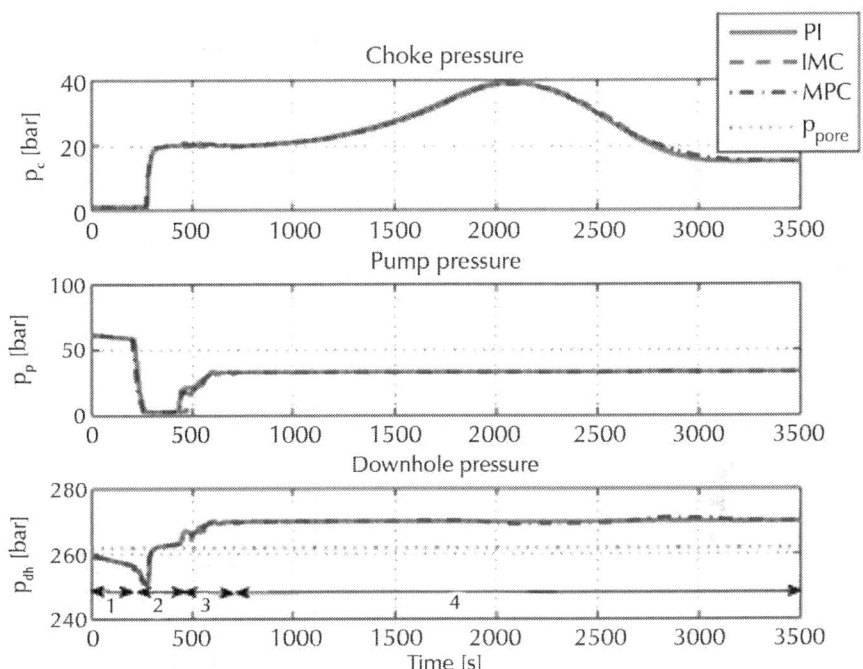

Figure 7: Simulation 1: choke pressure, pump pressure, and downhole pressure and pore pressure while taking a kick (1), shutting in the well (2), starting the pump (3) and circulating out the gas kick (4).

Table 6: Automatic well control simulation

Sequence	Description	Time period (s)
1	Influx	0–200
2	Shut in well	200–400
3	Start circulation	400–700
4	Circulate influx	700–3500

Fig. 7, Fig. 8, Fig. 9 and Fig. 10 show the results from simulations where the PI, IMC, and MPC controllers are used to control the pressure, plotted in solid green, dashed red and dashdotted blue, respectively. The pore pressure is shown in dotted magenta. The PI controller uses the choke opening to control the choke pressure or pump pressure, and the IMC and MPC controllers use both the choke opening and pump rate to control the pressure during the well control procedure.

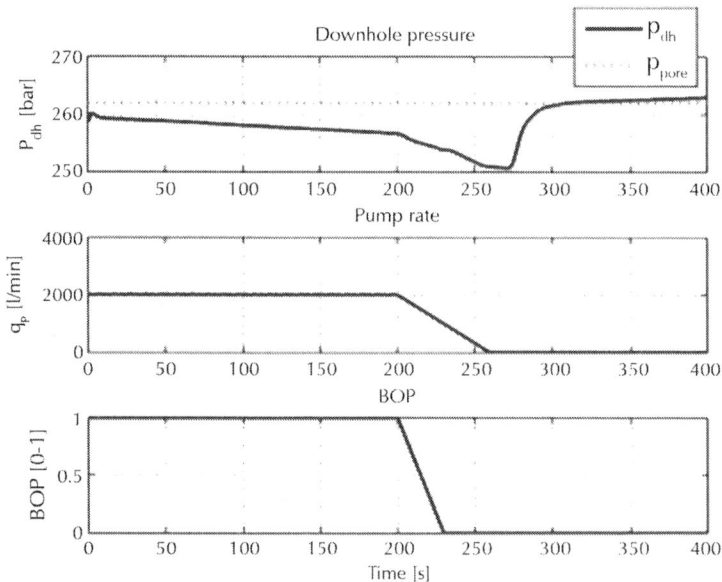

Figure 8: Simulation 1: taking a kick and shutting in the well. Figures from top to bottom: downhole pressure (solid blue) and pore pressure (dotted magenta), pump flowrate, and BOP opening. (For interpretation of the references to color in this figure legend, the reader is referred to the web version of the article.).

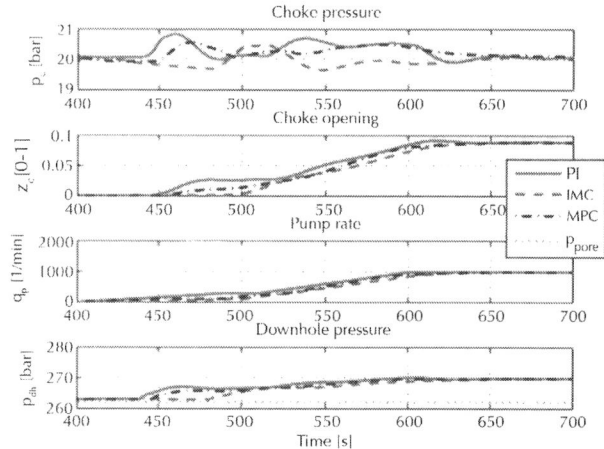

Figure 9: Simulation 1: choke pressure, choke opening, pump rate, and downhole pressure and pore pressure while starting up the pump after shut-in, controlling the choke pressure.

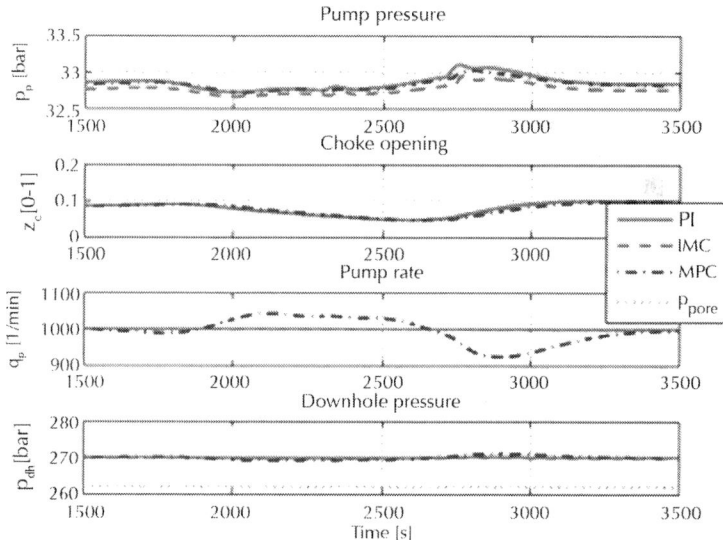

Figure 10: Simulation 1: pump pressure, choke opening, pump rate, and downhole pressure and pore pressure while the gas influx is circulating out of the choke, controlling the pump pressure.

The pit gain reaches the set limit of 2 m³ after about 200 s. The well control procedure then starts by shutting in the well by closing the BOP, shutting down the pump and closing the well control choke.

After the well has been shut in, the pressure at bottom hole increases, as depicted in the upper plot in Fig. 8 from 270 to 300 s. This is because as long as the pressure at bottom is below the pore pressure, gas will continue to flow into the closed well, which causes the pressure increase. The gas influx gradually reduces as the downhole pressure increases towards the pore pressure. When the downhole pressure has increased above the pore pressure the pressure stabilizes, as can be seen in Fig. 8 after 300 s.

When the influx is stopped and the pressure has stabilized, the next step in the well control procedure is to start up the pump in order to circulate out the gas that has entered the annulus. According to the procedure, the choke pressure should be kept constant during pump start-up. In the simulations, the automatic choke pressure controller (PI, IMC or MPC) is activated during pump startup, at time 400 s. The PI controller manipulates only the choke opening in order to control the pressure. The pump is then ramped up from 0 to 1000 l/min using sequence logic. The IMC and MPC controllers manipulate both the pump and the choke opening to control the choke pressure, with the target pump rate set to 1000 l/min. Fig. 9 shows the choke pressure, choke opening, pump rate, and downhole pressure and pore pressure during the pump startup sequence. The simulation using the PI controller is shown in solid green, the IMC in dashed red, and the MPC in dashdotted blue. All three controllers manage to control the choke pressure within ± 2 bar of the reference pressure. The reference pressure is the choke pressure at time 400 s, which is slightly different in the simulations of the different controllers.

When the pump is circulating at the target rate 1000 l/min after about 700 s, the procedure continues by circulating the kick out of the annulus. During this period, the pump pressure should be kept constant, so the controlled output is switched to pump pressure. Again, the PI controller manipulates only the choke opening, whereas the IMC and MPC use coordinated control of the choke opening and the pump rate. Fig. 10 shows the pump pressure, choke opening, pump rate, and downhole pressure during the time period when the gas exits the well control choke. All three controllers control the pump pressure within 0.5 bar of their reference pressure.

Simulation 2: Increased Reservoir Pore Pressure

To evaluate the robustness of the controllers, simulations have been run with a pore pressure of 267 bar instead of 262 bar in the reservoir section. The remaining simulation parameters are equivalent to Simulation 1. Results are shown in Fig. 11 and Fig. 12. The amount of gas entering the wellbore is doubled compared to when the pore pressure was 262 bar, when the time before the well is shut in remains the same.

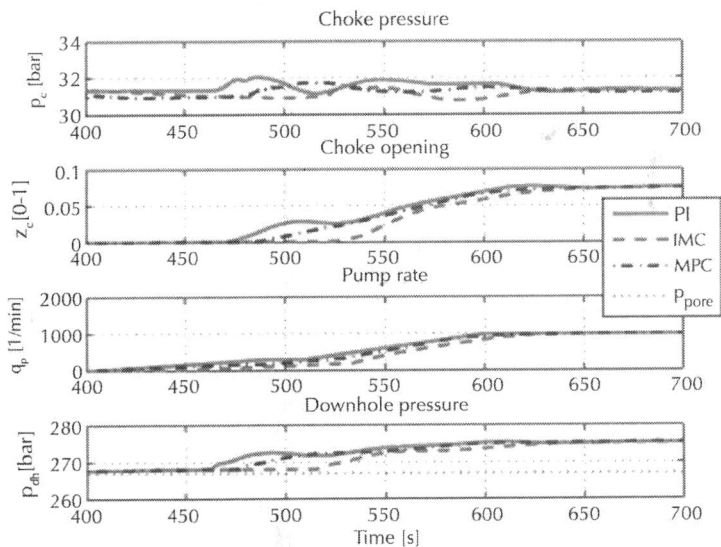

Figure 11: Simulation 2: choke pressure, choke opening, pump rate, and downhole pressure and pore pressure while starting up the pump after shut-in controlling the choke pressure.

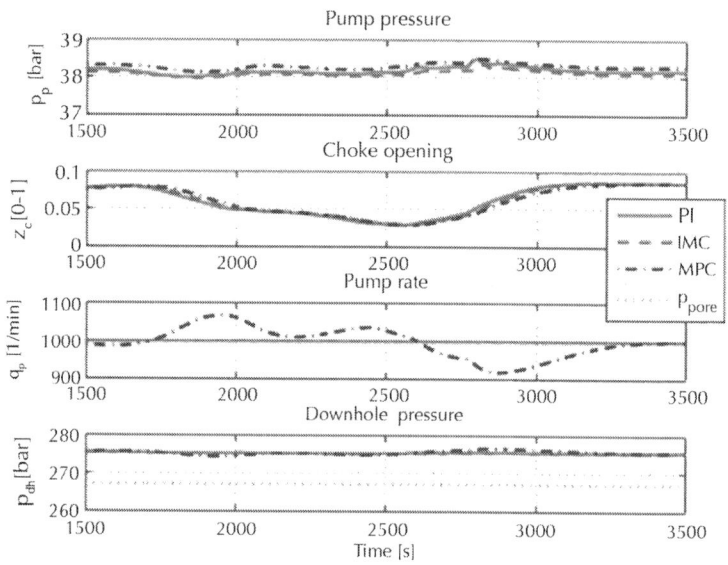

Figure 12: Simulation 2: pump pressure, choke opening, pump rate, and downhole pressure and pore pressure while the gas influx is circulating out of the choke controlling the pump pressure.

Fig. 11 shows the choke pressure, choke opening, pump rate, and downhole pressure and pore pressure during the pump startup sequence after the well has been shut in. The controlled output is the choke pressure, and the controllers are trying to maintain a constant choke pressure. All three controllers have the same or lower pressure variations compared to the previous simulation where the pore pressure was lower, even though there is more gas causing pressure disturbances in the system. This can be explained by the fact that since more gas has entered the well, the choke pressure after shut-in is higher during this simulation. When the pressure is higher, changes in choke opening will have a higher effect on the pressure, so the control response is faster and more aggressive in this case. This may also indicate a nonlinearity in the valve characteristics, and improved controller tuning could be investigated in a future study.

Fig. 12 shows the pump pressure, choke opening, pump rate, and downhole pressure and pore pressure at the time when the gas exits the choke. The controlled output is the pump pressure, and the controllers are trying to maintain a constant pump pressure. The reference pump pressure is higher during this simulation, since the downhole pressure must remain above the pore pressure, and the pore pressure is higher in this simulation. The controllers are a bit more aggressive in this case, both due to the higher pressure and that the higher amount of gas is causing more disturbances. All controllers are within the 2 bar pressure window.

Simulation 3: Increased Reservoir Permeability

To evaluate further the robustness of the controllers, simulations have been run where the permeability in the reservoir section is 400 mD instead of 100 mD, causing the gas to flow at a higher rate into the wellbore. The pore pressure is 262 bar, the same as in Simulation 1. The results are shown in Fig. 13 and Fig. 14.

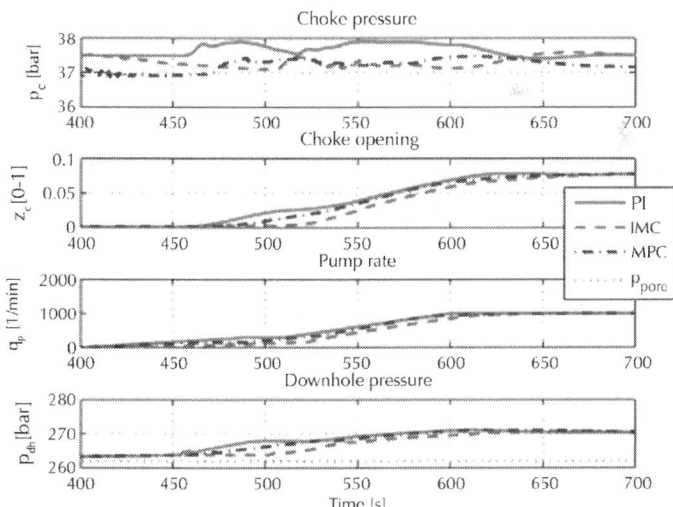

Figure 13: Simulation 3: choke pressure, choke opening, pump rate, and downhole pressure and pore pressure while starting up the pump after shut-in controlling the choke pressure.

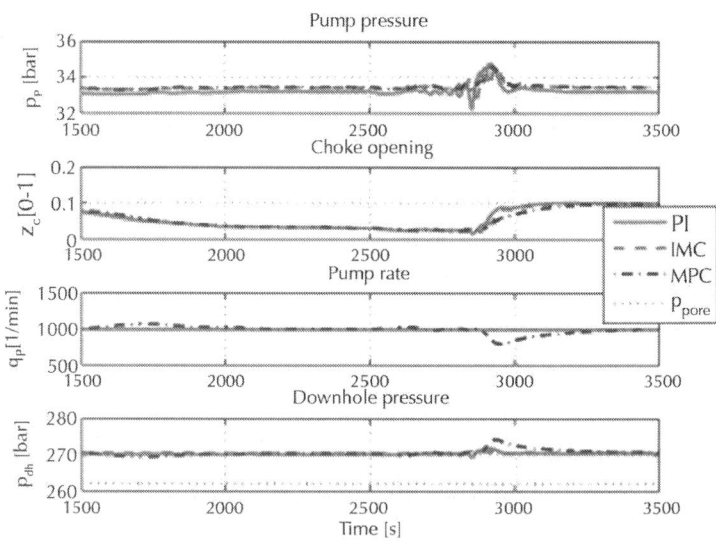

Figure 14: Simulation 3: pump pressure, choke opening, pump rate, and downhole pressure and pore pressure while the gas influx is circulating out of the choke controlling the pump pressure.

During this simulation the amount of gas entering the well is almost four times as high compared to Simulation 1 where the permeability was lower. Fig. 13 shows the choke pressure, choke opening, pump rate, and downhole pressure and pore pressure during the pump startup sequence after the well has been shut in. The controlled output is the choke pressure, and the controllers are trying to maintain a constant choke pressure. Also during these simulations the pressure variations are slightly less, compared to the simulation were the permeability was lower, even though there is more gas causing pressure disturbances in the system. Again, this can be explained by the fact that since more gas has entered the well, the choke pressure after shut-in is higher during this simulation, resulting in faster control response. But in this case the model based controllers are more aggressive, as tuned, causing more oscillations in the pressure. A further tuning of the possible nonlinearity in the valve characteristics could be investigated in a future study.

Fig. 14 shows the pump pressure, choke opening, pump rate, and downhole pressure and pore pressure at the time when the gas exits the choke. The controlled output is the pump pressure, and the controllers are trying to maintain a constant pump pressure. In this simulation, the reference pump pressure is almost the same as for Simulation 1, since the pore pressure is the same. The pressure variations are higher during these simulations for all the controllers, since there is much more gas causing disturbances in the pressure. But all the controllers stay well within the control criteria of 2 bar pressure variations.

Simulation 4: Well Control using Downhole Pressure Measurements

This section presents results from simulations where the downhole pressure measurements are available during the well control procedure. The pore pressure and permeability are the same as in Simulation 1. The simulation follows the same steps as Simulation 1, only the downhole pressure is now directly controlled and the PI, IMC and MPC controllers have the downhole pressure as controlled output during pump startup and while circulating out the kick.

Fig. 15 shows the downhole pressure and pore pressure, the choke opening, and the pump rate during the pump startup sequence. Fig. 16 shows the downhole pressure and pore pressure, the choke opening, and the pump rate while circulating out the kick. All the controllers manage to control the pressure within the 2 bar pressure window.

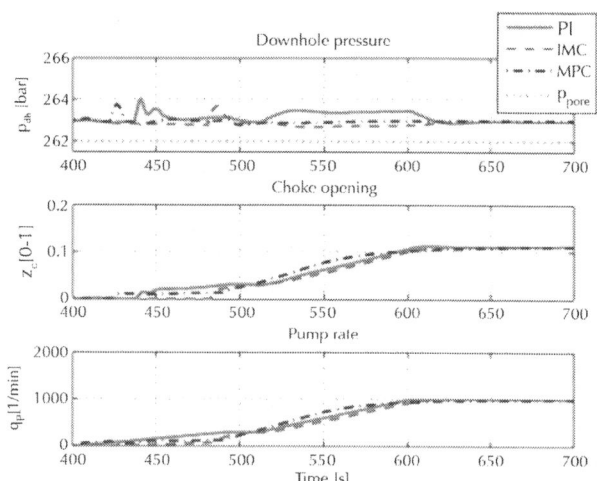

Figure 15: Simulation 4: downhole pressure and pore pressure, choke opening, and pump rate while starting up the pump after shut-in controlling the downhole pressure.

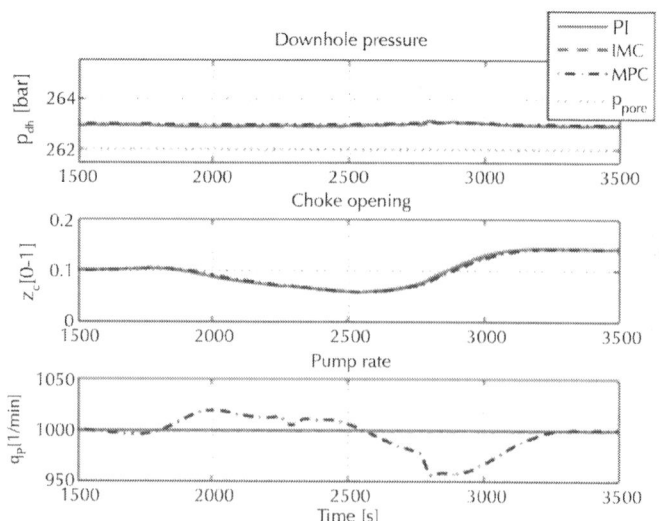

Figure 16: Simulation 4: downhole pressure and pore pressure, choke opening, and pump rate while the gas influx is circulating out of the choke controlling the downhole pressure.

Fig. 17 shows a comparison of the simulations with the IMC controller with and without downhole pressure measurement, during pump startup and while circulating out the kick. The controller using topside measurements is shown in dashed red, and the controller using downhole measurements is shown in dashdotted blue. During pump startup (from 400 to 700 s), the IMC using topside measurements controls the choke pressure. As can be seen from the dashed red line in the upper plot in Fig. 17, the downhole pressure then increases from 263 to 270 bar. The pressure increase is due to increased frictional pressure loss in the annulus when starting the pump and increasing the flowrate. This pressure increase is considered as an extra safety margin towards the pore pressure. When controlling the downhole pressure directly the pressure remains constant during pump start-up, as can be seen from the dashdotted blue line in the upper plot in Fig. 17.

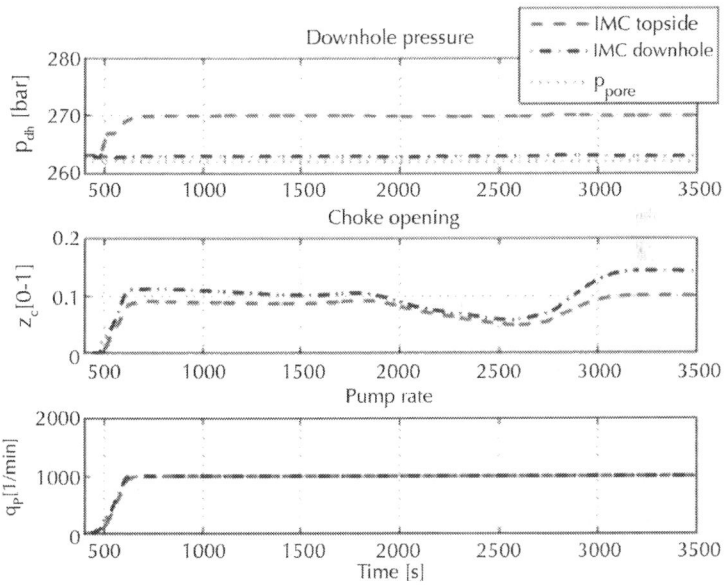

Figure 17: Comparison of IMC controller with and without downhole pressure measurement. From top to bottom: downhole pressure and pore pressure, choke opening and pump flowrate. Red dashed lines: simulation using the IMC controller with only topside measurements available

(choke pressure and pump pressure). Blue dashdotted lines: simulation using the IMC controller with downhole pressure measurements. (For interpretation of the references to color in this figure legend, the reader is referred to the web version of the article.).

To examine what happens when the controllers are less aggressive, a simulation was run with a pore pressure at 259 bar and where less gas enters the wellbore before the well is shut in. The IMC using topside measurements and the IMC using the downhole pressure measurement were evaluated for this case, and the results are shown in Fig. 18 during pump startup and while circulating out the kick. Both controllers manage to control the pressure even though the controllers are less aggressive in this case. During pump startup, a slower control response will give a higher margin towards the pore pressure since the choke will open more slowly when the pump is ramped up.

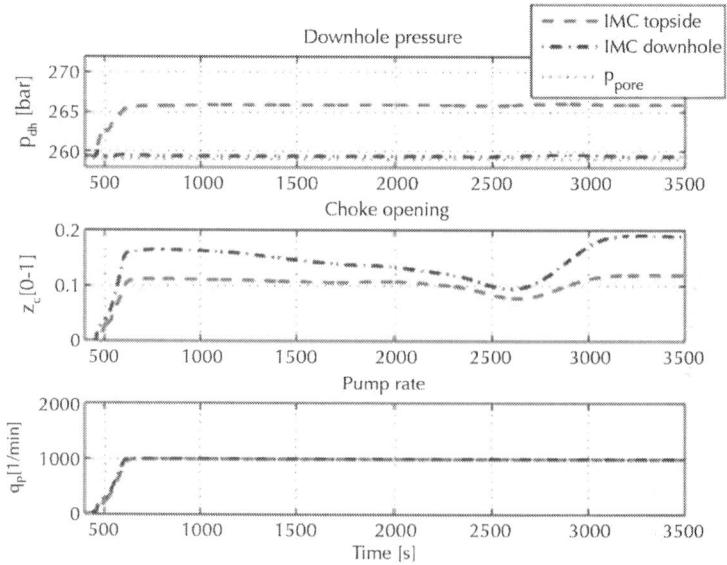

Figure 18: IMC controller with and without downhole pressure measurement when the pore pressure is 259 bar. From top to bottom: downhole pressure and pore pressure, choke opening and pump flowrate. Red dashed lines: simulation using the IMC controller with only topside mea-

surements available (choke pressure and pump pressure). Blue dashdotted lines: simulation using the IMC controller with downhole pressure measurements. (For interpretation of the references to color in this figure legend, the reader is referred to the web version of the article.).

DISCUSSION ON CONTROLLER TUNING

The various controllers used in the simulations have been tuned by looking at the dynamic response of the wellbore. The PI and the IMC controllers have few parameters and the selection of parameters has been relatively straight forward. Regarding the MPC controller, a larger set of parameters has been examined. A special focus has been on defining the best control horizon. Control performance improves by increasing the control horizon, but robustness is reduced as m becomes larger.

Fig. 19 shows the choke pressure during pump startup when using the MPC controller to control the pressure with various values for the control horizon m; m = 2, m = 4 and m = 6. The pore pressure is 262 bar, as in Simulation 1. Fig. 20 shows the results when running the simulation with a pore pressure at 267 bar, as in Simulation 2. In this case, there are more pressure oscillations with larger values for the control horizon m. With larger control horizon, the MPC controller tries to be as close as possible to the inverse of the built-in linear model. When the pore pressure is higher, more gas has entered the wellbore before shut-in. The linear model in the MPC controller is not accurate with the large amount of gas, and the resulting control action is too aggressive, thus creating the observed oscillations. We have found the optimal control horizon in this case to be m = 2.

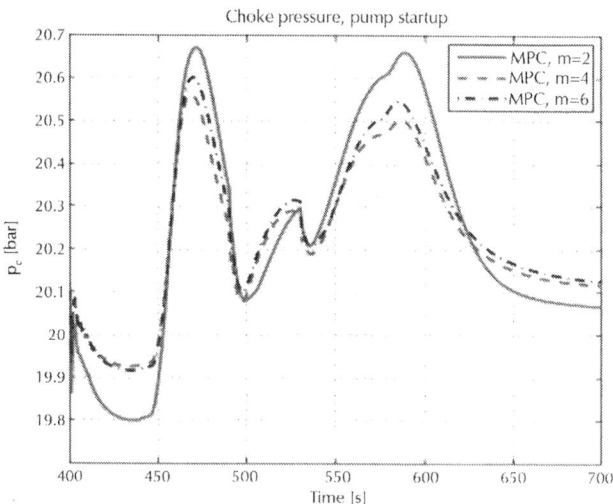

Figure 19: MPC tuning, control horizon: choke pressure while starting up the pump after shut-in controlling the choke pressure using MPC control with various control horizon values m. Pore pressure at 262 bar.

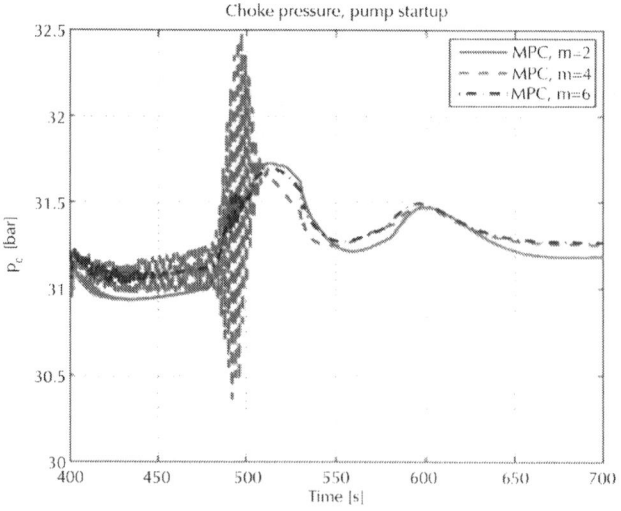

Figure 20: MPC tuning, control horizon: choke pressure while starting up the pump after shut-in controlling the choke pressure using MPC control with various control horizon values m. Pore pressure at 267 bar.

CONCLUSIONS AND FURTHER WORK

The simulations show that it is feasible to control the pressure using automatic control of the choke valve and pump during an unexpected gas influx by use of the presented control methods. The control methods are robust against changes in process conditions and disturbances, as they are able to handle several pressure levels and gas volumes without requiring re-tuning. However, since the pressure dynamics in the well is influenced when gas is entering the well, the model based controllers could probably be improved if the models were updated after the gas influx occurred.

The results indicate that adaption of the automatic sequence to the current manual procedure is applicable. However, to avoid a reduction in downhole pressure when stopping the pump and shutting in the well, the automatic sequence may be further improved beyond what is feasible with manual operation.

The current methods show that automatic pressure control while gas is present in the well is possible to achieve. Underbalanced drilling (UBD), where the well pressure is below the pore pressure during the whole operation, is closely related to the well control operation. The pressure control during UBD operations is currently performed manually, but this operation could be automated. The path towards automation of the UBD operation and the well control operation should be followed in close relation.

Two different paths should be followed regarding the future work on the topic on automatic well control for drilling operations. Firstly, there should be an investigation how the simplified models may be updated after the gas influx occurs. This could lead to a further improved performance of the model based controllers. The pore pressure and permeability range within which the controllers are robust could also be examined more thoroughly. In the simulations it was only examined that the downhole pressure did not fall below the pore pressure while circulating the kick. The upper pressure constraint is the fracture pressure of the formation at the casing shoe, which could be investigated further in future work. The second path that should be followed regarding future work is the shut-in sequence. The current shut-in sequence is only using sequence logic and gives reduced downhole pressure. From the simulations we see that this leads to increased influx. Therefore, an improved shut-in control sequence utilizing both pump flow rate control and choke valve pressure control should be investigated.

ACKNOWLEDGMENTS

The authors acknowledge the Research Council of Norway for grant number 176018 of the PETROMAKS program and also ConocoPhillips and Statoil for additional funding. Thanks also to Dr. Erlend H. Vefring for useful comments.

REFERENCES

1.   R.D. Grace, Blowout and Well Control Handbook, Elsevier Science, Burlington, MA, 2003.

2.   K. Malloy, C. Stone, J.G.H. Medley, D. Hannegan, O. Coker, D. Reitsma, H. Santos, J. Kinder, J. Eck-Olsen, J. McCaskill, J. May, K. Smith, P. Sonneman, Managedpressure drilling: what it is and what it is not, in: IADC/SPE Managed Pressure Drilling and Underbalanced Operations Conference and Exhibition, San Antonio, TX, 2009.

3.   D. Reitsma, Development and application of combining a real-time hydraulics model and automated choke to maintain a relatively constant bottomhole pressure while drilling, in: International Petroleum Technology Conference, Doha, Qatar, 2005.

4.   D. Reitsma, T. Runggai, N. Hudson, R. Zaeper, O. Backhaus, M. Hernandez, Successful implementation of first closed loop, multiservice control system for automated pressure management in a shallow gas well offshore Myanmar, in:

SPE/IADC Drilling Conference, Amsterdam, The Netherlands, 2008.

5.   J. Godhavn, Control requirements for high-end automatic mpd operations, in: SPE/IADC Drilling Conference and Exhibition, Amsterdam, The Netherlands, 2009.

6.   J. Zhou, G. Nygaard, Control and estimation of downhole pressure in managed pressure drilling operations, in: Proceedings of the 4th International Symposium on Communications, Control and Signal Processing, Limassal, Cyprus, 2010. 316 L.A. Carlsen et al. / Journal of Process Control 23 (2013) 306–316

7.   J.-M. Godhavn, A. Pavlov, G.-O. Kaasa, N.L. Rolland, Drilling seeking automatic control solutions, in: Proceedings of the 18th IFAC World Congress, Milano, Italy, 2011.

8.   Ø. Breyholtz, G. Nygaard, M. Nikolaou, Advanced pressure control for dualgradient drilling, in: SPE Annual Technical Conference and Exhibition, New Orleans, LA, 2009.

9.   G. Nygaard, G. Nævdal, Non-linear model predictive control scheme for stabilizing annulus pressure during oil well drilling, Journal of Process Control 16 (7) (2006) 719–732.

10.   Ø. Breyholtz, G. Nygaard, M. Nikolaou, Managed pressure drilling: a multi-level control approach, in: SPE Intelligent Energy Conference and Exhibition, Utrecht, The Netherlands, 2010.

11.   Ø. Breyholtz, G. Nygaard, M. Nikolaou, Automatic control of managed pressure drilling, in: Proceedings of American Control Conference, Baltimore, USA, 2010.

12.   M. Davoudi, J. Smith, B. Patel, J. Chirinos, Evaluation of alternative initial responses to kicks taken during managed pressure drilling, in: IADC/SPE Drilling Conference and Exhibition, New Orleans, LA, USA, 2010.

13.   H. Santos, P. Reid, J. Jonesand, J. McCaskill, Developing the micro-flux control method-part 1: system development field test preparation and results, in: SPE/IADC Middle East Drilling

Technology Conference and Exhibition, Dubai, United Arab Emirates, 2005.

14. M. Stanislawek, J.R. Smith, Analysis of alternative well-control methods for dual-density deepwater, in: IADC/SPE Drilling Conference, Miami, FL, USA, 2006.

15. J. Zhou, G. Nygaard, Automatic model-based control scheme for stabilizing pressure during dual-gradient drilling, Journal of Process Control 21 (8) (2011) 1138–1147.

16. J. Zhou, G. Nygaard, J. Godhavn, Ø. Breyholtz, E. Vefring, Adaptive observer for kick detection and switched control for bottomhole pressure regulation and kick attenuation during managed pressure drilling, in: Proceedings ofAmerican Control Conference, Baltimore, USA, 2010.

17. J. Gravdal, M. Nikolaou, Ø. Breyholtz, L. Carlsen, and Improved kick management during mpd by real-time pore-pressure estimation, in: SPE Annual Technical Conference and Exhibition, New Orleans, LA, 2009.

18. R.J. Lorentzen, G. Nævdal, A. Lage, Tuning of parameters in a two-phase flow model using an ensemble kalman filter, International Journal of Multiphase Flow 29 (2003) 1284–1309.

19. E. Cayeux, B. Daireaux, Early detection of drilling conditions deterioration using real-time calibration of computer models: field example from North Sea drilling operations, in: Proceedings ofthe SPE/IADC Drilling Conference and Exhibition, Amsterdam, The Netherlands, 2009.

20. E. Cayeux, B. Daireaux, E. Dvergsnes, G. Sælevik, M. Zidan, An early warning system for identifying drilling problems: an example from a problematic drillout cement operation in the north-sea, in: Proceedings of the SPE/IADC Drilling Conference and Exhibition, San Diego, CA, 2012.

21. G.-O. Kaasa, A Simple Dynamic Model of Drilling for Control, Tech. rep., StatoilHydro, Porsgrunn, Norway, 2007.

22. Ø.N. Stamnes, J. Zhou, G. Kaasa, O. Aamo, Adaptive observer design for the bottomhole pressure of a managed pressure

drilling system, in: Proceedings of the 47th IEEE Conference on Decition and Control, Cancun, Mexico, 2008.

23. M. Morari, E. Zafiriou, Robust Process Control, Prentice-Hall, USA, 1989.

24. J. Maciejowski, Predictive Control with Constraints, Pearson Education, Essex, England, 2002.

Chapter 4

Potential Implementation of Underbalanced Drilling Technique in Egyptian Oil Fields

K.A. Fattah[a], S.M. El-Katatney[b], and A.A. Dahab[b]

[a]Petroleum and Natural Gas Engineering Department, College of Engineering, King Saud University, , Riyadh 11421, Saudi Arabia
[b]Petroleum Engineering Department, Faculty of Engineering, Cairo University, Egypt

ABSTRACT

The need to increase productivity and to reduce drilling damage favors the use of underbalanced drilling (UBD) technology. In highly depleted reservoirs, extremely low-density fluids, such as foams or aerated mud, are used to achieve circulating densities lower than the pore pressure. In such cases, the induced modification of the in situ stresses has to be supported mainly by the rock, with little contribution from the drilling fluid pressure. The application of underbalanced drilling depends on the mechanical stability of the drilled formation, among other factors. In general, poorly consolidated, depleted formations are not suited for that technology.

In this paper, 23 UBD worldwide cases have been analyzed; two of which are from Egyptian fields and the others are from Iran, Algeria, Kuwait, Oman, Texas, Mexico, Indonesia, Canada, Libya, Middle East, Qatar, Saudi Arabia and Lithuania. From these analyses, the reasons of failure or success have been stated. The reasons of success included depleted reservoirs and highly fractured carbonates formation while, the reasons of failure include over pressurized shale, highly tectonic stress areas, and downhole failures. The main attractive application of this technology was proposed to be only in the reservoir section, and the target was to prevent the reservoir damage and hence increase the productivity and recovery factor.

A proposed underbalanced drilling program is developed based on these analyses to be used in the three main regions in oil and gas producing Egyptian fields. The aerated mud was selected as a drilling fluid to drill the reservoir section in Western Desert and Gulf of Suez region whereas the single phase fluid was selected as a drilling fluid in the Nile Delta region.

INTRODUCTION

Drilling cost is considered one of the major components of operating cost in the petroleum industry. Improving the penetration rate of drilling operation and reducing drilling problems, such as pressure differential pipe sticking and lost circulation, have long been considered an effective way of decreasing drilling costs. The overbalance pressure, generally recognized as the most important among the many factors affecting penetration rate, is often defined as the pressure differential between the borehole pressure and formation fluid pressure (Murray and Cunningham, 1955, Eckel, 1957, Cunningham and Eenink, 1959, Gamier and van Lingen, 1959, Vidrine and Benit, 1968, Bourgoyne and Young, 1974a, Bourgoyne and Young, 1974b and Black and Green, 1978). Formation pressures lower than the static pressure of a column of fresh water require the use of a lighter fluid, such as air, injected with liquid to obtain lower overbalance pressure to enhance penetration rate and to minimize lost circulation and pipe sticking as well as formation damage. Therefore, aerated mud drilling "implies the use of air or natural gas as the circulating medium instead of the regular mud" is becoming an attractive practice in some areas. The commercial use of aerated mud drilling began only in recent years (Rankin et al., 1989 and Claytor et al., 1991). Low-density drilling fluids used in underbalanced drilling consist of air, mist, stable foam, and aerated mud foam with back pressure. Whereas the term "aerated mud" implies the simultaneous introduction of air and mud together into the standpipe in order to drill special types of formations (Godwin et al., 1986, Boyun and Rajtar, 1995 and Salah El-Din and El-Katatney, 2009).

The main advantage of air as a circulating fluid is that being the lowest density fluid. It imposes minimum pressure on the formation to be drilled. High penetration rates have been achieved in hard and dry formations with the use of air as a circulating fluid. In addition to high penetration rate, longer bit life results through the use of this medium as compared to mud. Drilling rates as high as 90 ft/h have been attained in shales. Air drilling, however, is restricted to areas where high volume water sands are not present ahead of the producing zone. The rate of water influx that can be handled in the case of air drilling is also not well known. Other inherent disadvantages of using air or natural gas as drilling fluids include possibility of downhole fires and explosions, and sloughing of formations due to underbalance of stresses around the wellbore. Possibility of downhole explosions are of particular concern in air drilling operations. Small dust-like particles are generated as a result of rock cuttings (chips) being ground and pulverized by the drill string in the annulus, and collision of cuttings with each other, the tool joints, and the wall of the borehole due to the high velocity forces. In the presence of moisture, seal rings may form at tight places in the annulus, which create pressure chambers. With additional influx of natural gas from gas-bearing zones being penetrated by the bit, an explosion may easily occur.

Besides having formations suitable for air drilling, the most important consideration in drilling with air is the volume of air required. Air drilling often fails because of insufficient volume of air necessary to clean the hole efficiently under certain conditions, e.g., wet hole, sloughing shales, and influx of formation water. A practical rule of thumb for determining adequate air volume is that the volume required achieving 1000 ft per minute annular velocity to clean the hole properly (Godwin et al., 1986 and Boyun and Rajtar, 1995).

Drilling with foam has some appeal due to the fact that foam has some attractive qualities and properties with respect to the very low hydrostatic densities, which can be generated with foam systems (Hooshmandkoochi et al., 2007, Moore and Lafave, 1956, Maurer, 1998 and Bentsen and Veny, 1976). Foam has good rheology and excellent cutting transport properties. The fact that foam has some natural inherent viscosity as well as fluid loss control properties, which may inhibit fluid losses, makes foam a very attractive drilling medium. During connections and trips, the foam remains stable and provides a more stable bottom hole pressure. It is a particularly good drilling fluid with a high carrying capacity and a low density. The foam normally remains stable, even when it returns to the surface, and this can cause problems on a rig if the foam cannot be broken down fast enough. In earlier foam systems, the amount of defoamer had to be tested carefully so that the foam was broken down before any fluid entered the separators. In closed circulation drilling systems, stable foam could cause particular problems with carry over. The recently developed stable foam systems are simpler to break, and the liquid can also be refoamed so that less foaming agent is required and a closed circulation system can be used. These systems, in general, rely on either a chemical method of breaking and making the foam, or the utilization of an increase and decrease of pH to make and break the foam. The foam quality at surface used for drilling is normally between 80% and 95%. The quality of foam means that the system is 80–95% gas, with the remaining 5–20% being liquid. Downhole, due to the hydrostatic pressure of the annular column, this ratio changes as the volume of gas is reduced. An average acceptable bottom-hole foam quality (FQ) is in the region of 50–60%. Fluid densities for foam range from 1.6 ppg to 6.95 ppg (0.2–0.8 S.G.) (Godwin et al., 1986 and Boyun and Rajtar, 1995). The density ranges are adjusted with the make up of the foam by adjusting the Liquid Volume Fraction (LVF) through the injection of liquid and gas by adjusting the backpressure on the well. The backpressure adjusts the downhole pressure and slows down the velocities in the annulus. Experience has proven that foam is able to handle over 100 bbl/h of water influx (Godwin et al., 1986, George and Waston, 1956 and Boyun and Rajtar, 1995).

So, the objective of this research work is to investigate and analyze many worldwide applications of underbalanced drilling and state the reasons of success or failure of this application. Based on these analyses, a proposed underbalanced drilling program is developed. In this proposed program, the method of selecting the appropriate technique to be applied for these candidate areas are selected according to the geology of the area and the bottom hole conditions inside the wells.

STUDIED CASES

In this section, three case studies from Egyptian fields and other places are analyzed in detail and a summary of 20 cases from other worldwide fields are given with a brief discussion about their objectives, problems and results (Salah El-Din and El-Katatney, 2009).

Case 1: Gulf of Suez Area

The well is located at onshore Belayim oil field. The well target was sandstone of zone III (Belayim formation, Feiran member) at a total depth of 2335 m TVD, 2854 m MD. The pressure in Zone III (sandstone) was estimated to be 3000–3500 psi (0.3917–0.4569 psi/ft). The objectives of UBD were to increase rate of penetration, enhance Well control, reduce occurrence of lost time incidents, and increase well productivity. The 20 m of the new hole at 7 in. liner shoe at 2659 m MD was drilled with only mud, then the MWD signal test was performed (inflow test and also to test the optimum rate combination for better MWD signal) as shown in Table 1. Based on this test, the formation pressure was estimated to be less than 2500 psi that was confirmed at 2400 psi from vacuum test and the MWD can work up to 21% nitrogen. Nitrified mud (500 SCFM + 230 gpm diesel) was applied while close balance drilling the 6 in. original and side-track lateral section. The 6 in. hole was drilled to depth 2830 m utilizing UBDS and powerpack motor of 1.15° BH c/w MWD Impulse, VPWD, ADN tools (inclination at bit, annulus and string pressure, GR resistivity, density-neutron) with 2 × 3-1/2 in. W.FORD float valve + motor restriction sub (nozzle 14/32 in.) for improving MWD signal. The analysis of this well results showed that, The ROP was enhanced drastically in sand from 4 m/h while sliding to 50 m/h, and in anhydrite was 8–10 m/h (experienced 2–4 m/h in normal overbalance drilling), the use of rotating head helped to control well while tripping and also in case of separator carry over problems, and the Crew acquired UBD work experience.

Table 1: Change in BHCP versus mud rate and N2 rate

Duration (h)	Mud rate (GPM)	N$_2$ rate (SCF/m)	N$_2$ (%)	SPP (psi)	BHCP (psi)	ECD (kg/lit)	Gain (bbls)	MWD Signal
1.5	250	250	4.8	2250	2887	0.86	31 mud	Ok
3.0	250	500	9.1	2250	2660	0.79	34 mud	Ok

2.0	240	500	11.4	1800	2652	0.79	0	Ok
1.5	230	500	13.2	1600	2620	0.78	0	Ok
1.5	210	500	15.5	1450	2590	0.77	0	Ok
1.5	180	500	21.6	1120	2549	0.76	0	Ok

Case 2: Western Desert Gas Field Area

The well is located at the central part of the western desert block. The well target was to drill 3-7/8 in. × 500 m horizontal section in unit 3 of the Mesozoic Lower Safa reservoir. They are composed of low to medium permeable (1–500 md) micaceous sandstones deposited in a strong tidally influenced estuary,Fig. 1. Lower Safa formation comprises a high-energy sequence of Estuarine deposits with a total average thickness of 110 m in the area where is planned, although only 29 m of these thickness are considered productive. The objective of UBD was to prevent reservoir damage. Gasification was through drill pipe injection technique.

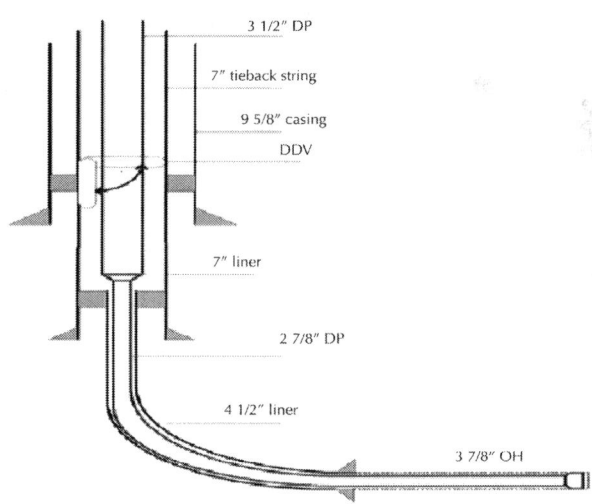

Figure 1: Well profile diagram for case 2: western desert gas field area.

The well was completed as open hole. Average ROP during overbalanced drilling operations on offset wells has been historically 2–3 m/h in the horizontal section. Historical data for UBD wells suggested that there will be an improvement in ROP due to the elimination of the chip hold-down effect. It was estimated that the ROP will be between 5 and 10 m/h. The drilling fluid of choice was produced water. The drilling fluid could be separated from the produced hydrocarbons and re-used. Due to the CO_2 content of the reservoir (up to 9%) and the use of nitrogen (up to 5% O2), corrosion mitigation was required. Once the well started to produce during the drilling phase, the N_2 was stopped, which in turns eliminated excessive use of corrosion inhibitors. Water and nitrogen gave the desired underbalanced margin when kicking off the well, and water was treated with suitable chemicals for corrosion mitigation. It became apparent that the Lower Safa formation was normally pressured. Hence by using just water, the BHP will be 260 psi underbalanced. Nitrogen was required to create a greater draw down than the 260 psi as it is unknown at what draw down the matrix starts to contribute to the inflow.

As soon as the well produced, nitrogen was cut down to zero rates. Nitrogen injection was required again every time the drill string tripped through the Down-hole Deployment Valve (DDV) to remove the water from the reservoir section.

Fig. 2 shows the working window (operating envelope) for the well (case 2) with no reservoir inflow for, 3-7/8 in. hole, 3-1/2 in. × 2-7/8 in. drill pipe design, 2 × 500 m legs, and bit at TD. Also plotted on the operating envelope, are the various constraints that must be fulfilled during underbalanced drilling operations. After drilling 200 m, the drilling had been stopped due to failure of downhole equipment due to high temperature.

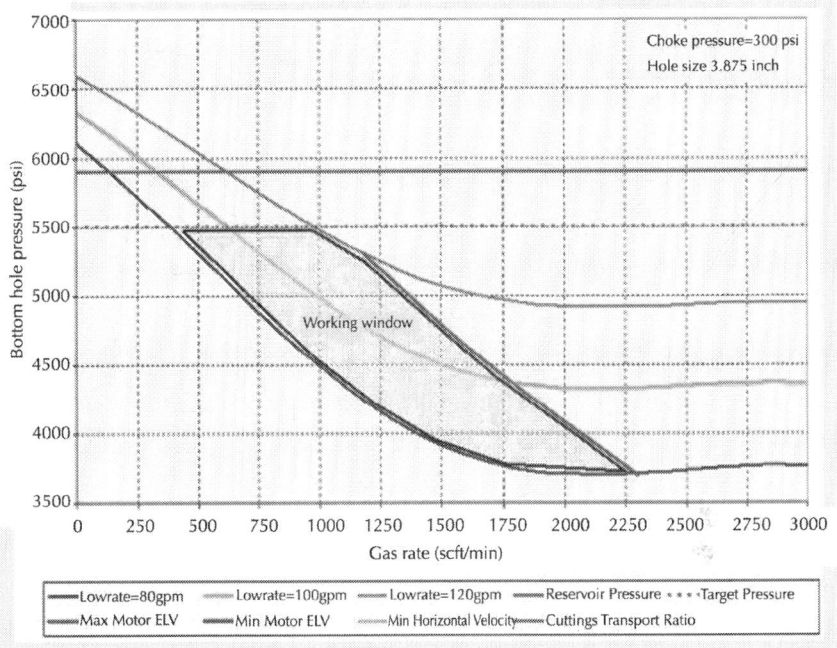

Figure 2: Working window for case 2: western desert gas field area.

Case 3: Iranian Oil Field

The target reservoir for this well was Asmari formation, the formation was fractured carbonated formation. The reservoir drive mechanism was gas cap. Shale strings were not expected in this formation. Expected reservoir pressure and temperature were 2622 psi and 141 °F, respectively. Reservoir fluid was oil with API gravity of 25°, GOR 564 SCF/STB, and H2S concentration of 240 ppm. The permeability of the reservoir was 0.1–1000 md with a porosity of 9% (Hooshmandkoochi et al., 2007). The well was drilled from m (9-5/8 in shoe depth) to a total depth of 2938 m MD (2567 m TVD), Fig. 3. The primary objectives of this underbalanced drilling project were to: minimize drilling induced formation damage, eliminate drilling fluid losses, and improve drilling performance. The drilling fluid selection was one of the most critical decisions in planning an underbalanced well. The right fluid(s) selection will not only lead to suitable BHCP but will also minimize pressure transients and thus eliminating/minimizing formation impairment. The deviated underbalanced section of this well was to be drilled with a Gachsaran field native crude oil and a membrane nitrogen generation circulating system. Liquid Phase, the native crude oil, was chosen over Diesel and other drilling fluids because it is the natural reservoir fluid for this well. This minimized chances of formation damage in event of pressure transients and/or from fluid imbibitions. The well was displaced with the produced fluid after getting enough oil production. Gas Phase, nitrogen, was selected as the injection gas because of its inert nature, economic availability and suitability for this specific underbalanced drilling project. Nitrogen was obtained from the surrounding air and generated onsite, by nitrogen production unit (NIOC's). The multiphase flow behavior in the wellbore during underbalanced drilling was very complex. The response of the downhole conditions to changes in various flow parameters must be characterized prior to the commencement of underbalanced drilling operations in order to maximize chances of success.Fig. 4 contains a plot of the bottom hole circulating pressures induced by a variety of nitrogen rates and the Gachsaran native crude oil injection rates. This plot was referred to as the operating envelope. Also plotted on the operating envelope, are the various constraints that must be fulfilled during underbalanced drilling operations. The range of flow rates that satisfy all of the constraints, defined the acceptable operating region. A minimum drawdown at the bit of 200 psi was required to ensure adequate underbalanced conditions in the well, with a maximum drawdown of 300 psi to minimize any near wellbore depletion effects. The target bottom hole circulating pressure at the bit for this well was 2300–2400 psi.

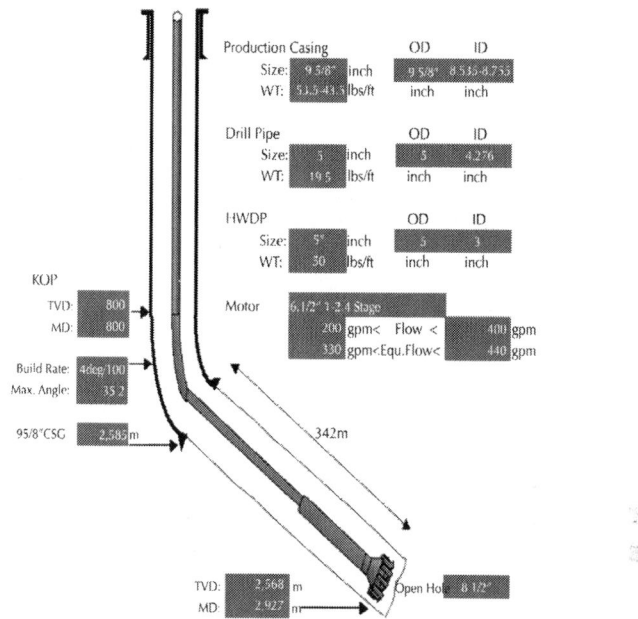

Figure 3: Well profile diagram for case 3: Iranian oil field area.

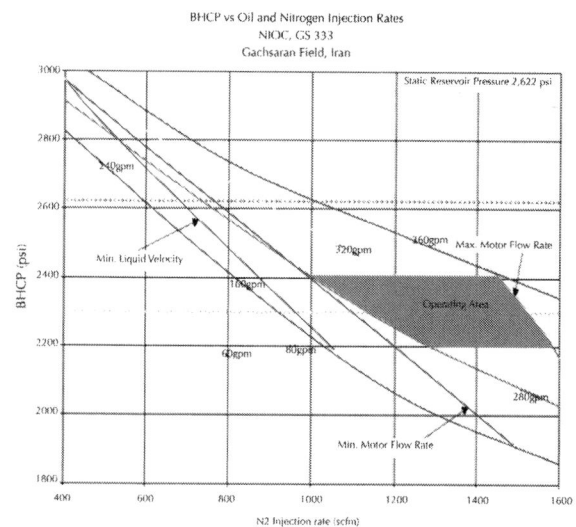

Figure 4: Operational envelope – native crude for case 3: Iranian oil field area.

UBD on this well experienced some typical logistical and start up problems associated with a steep learning curve, this being the first such operation in Iran. Despite all the problems encountered in this well, the following performance had been achieved: drilled to 308 m of total open hole depth, no loss circulation was encountered while drilling, successfully implemented UBD technology, and no Quality, health, safety and environment (QHSE) incidents were recorded. Data for case 4 to case 23 are given in the (Azeemddin, 2006, Bates, 1965, Bennion et al., 1998, Dorenbos and Ranalho, 2002, Gordon, 2005, Gray, 1957, Hongren et al., 1999, International Association of Drilling Contractor, 2005, Kuru, 1999, Louison et al., 1984, Maclovio, 1996, Meng, 2005, Moore et al., 2004, Nas, 2004, Negra et al., 1999, Parra et al., 2003,Qutob, 2007, Qutob and Ferreira, 2005, Sunthankar, 2001, Weatherford Company, 2006, Westermark, 1986, Whiteley and England, 1986 and Zhou, 2005).

Data Analysis

The following analysis is carried out based on some actual wells drilled underbalanced worldwide. As mentioned before, the main advantage of underbalanced drilling techniques is to increase the rate of penetration as compared with overbalanced drilling techniques.

Table 2 gives the recorded data that were collected from successful underbalanced drilling cases in which the aerated mud was used to drill sandstone reservoir sections (Moore and Lafave, 1956).

Table 2: Recorded ROP in Algeria

Algeria sandstone reservoir		
Well number	ROP overbalanced (ft/h)	ROP underbalanced (ft/h)
1	10.4	19.5
2	10.4	17.6

3	19.3	22.5
4	19.5	22.3
5	13.5	45
6	17	26.6

From Fig. 5, there is an observed increase in ROP in all cases that were drilled by underbalanced techniques. In underbalanced drilling, ROP was increased due to the disappearance of chip hold-down effect. So the normal trend includes that an increase of the ROP resulted from a decrease in the hydrostatic pressure of drilling fluid as compared with the pressure of the formation that drilled by UB, as shown in Fig. 6.

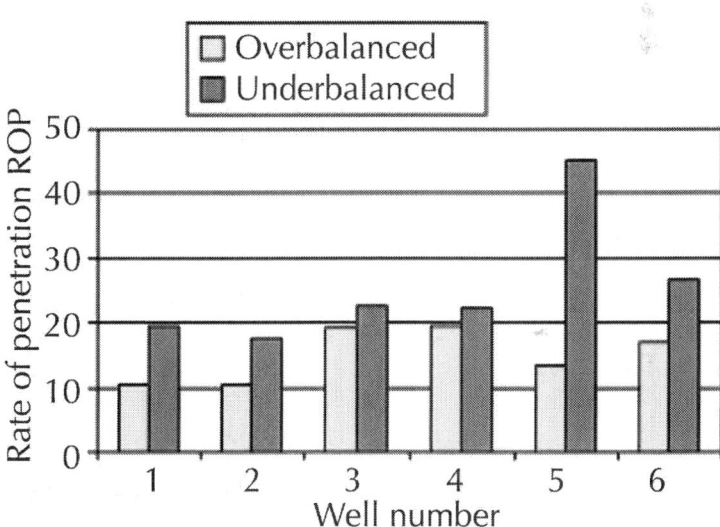

Figure 5: Comparison between ROP in OB and UB cases.

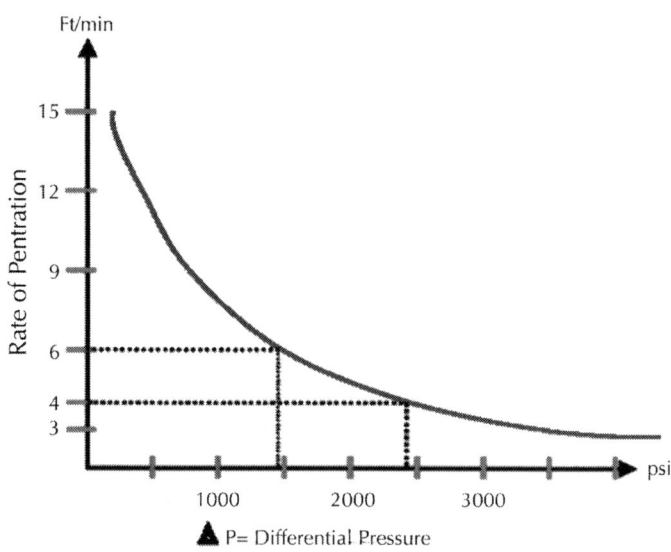

Figure 6: Relationship between ROP and pressure drop.

Table 3 gives the recorded data of ROP (ft/h) and pressure drop (psi) for different reservoirs that were drilled by aerated fluid as an UBD drilling fluid. These reservoirs have the same lithology but having different reservoir pressure.

Table 3: ROP versus pressure drop for UBD wells

Reservoir pressure (psi)	Pressure drop (ΔP) (psi)	Rate of penetration (ft/h)	Lithology
2900	290	45	Sandstone
3000	360	38	Sandstone
1350	540	16	Sandstone
3200	640	27	Sandstone
5500	990	30	Sandstone

Table 4 gives a recorded data for different wells drilled by aerated fluid in a reservoir that has a constant pressure and same lithology compared to those wells drilled in overbalanced environment (Moore and Lafave, 1956).

Table 4: Recorded data for UBD wells

Pressure drop (psi)	ROP (ft/h)	Production while drilling (%)	Production after test (%)	Lithology
290	26.6	0	1.2	Sandstone
320	44.7	0.8	3.9	Sandstone
350	19.45	1	2	Sandstone
406	22.5	1.5	1.8	Sandstone
435	17.6	2.7	3.4	Sandstone

Fig. 7 illustrates that ROP initially decreases with an increase in pressure drop and increases with further increase in pressure drop. Whereas, Fig. 8 shows that ROP has no definite relation with pressure drop if other drilling parameters are ignored. However a continuous increase in formation fluid production while drilling was observed with the continuous increase in pressure drop as shown in Fig. 9.

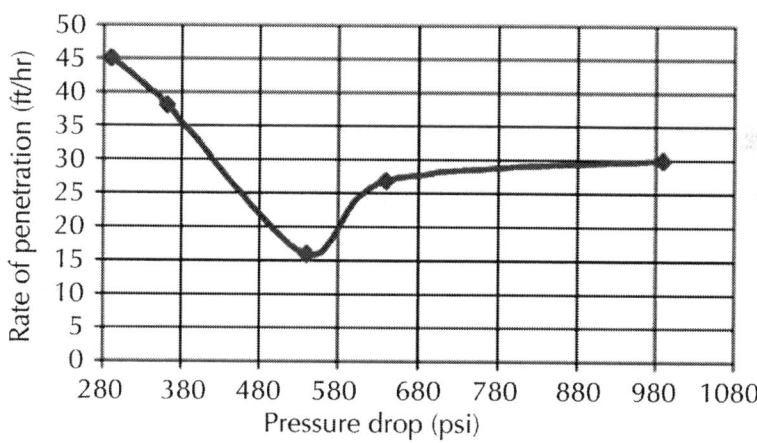

Figure 7: ROP versus pressure drop for UBD wells in different reservoirs.

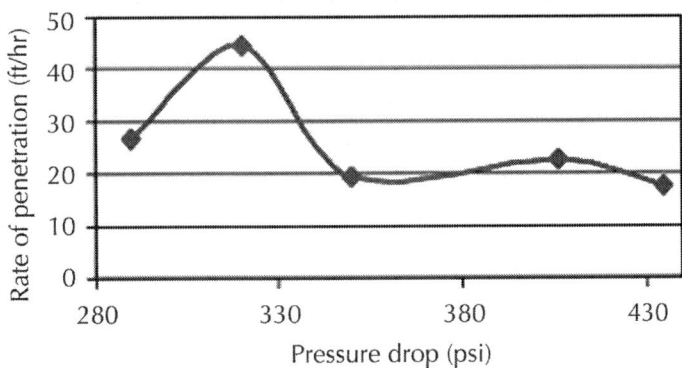

Figure 8: ROP versus pressure drop for UBD wells in one reservoir.

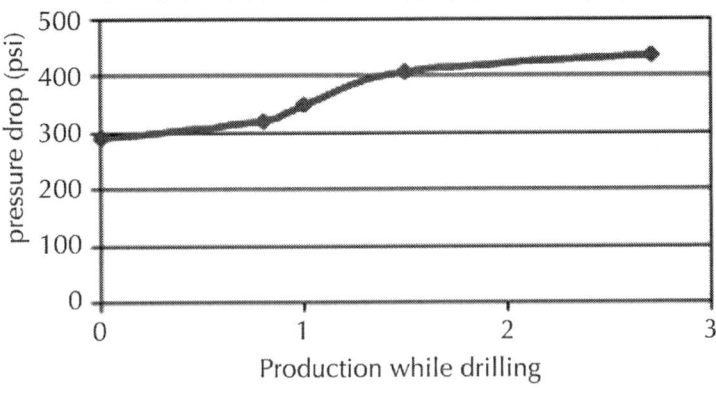

Figure 9: Production while drilling versus pressure drop for UBD wells.

Fig. 10 illustrated that all wells drilled by UBD have an increased in fluid production rate compared to those wells drilled in overbalanced environment. In addition, there is no clear relation between the amount of fluid production while drilling and the amount of fluid production after the well is put on production as shown inFig. 10.

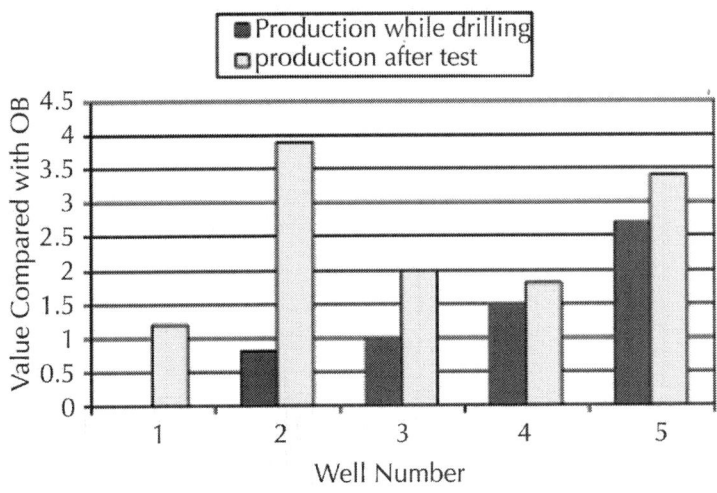

Figure 10: Comparison of production while and after UBD drilling.

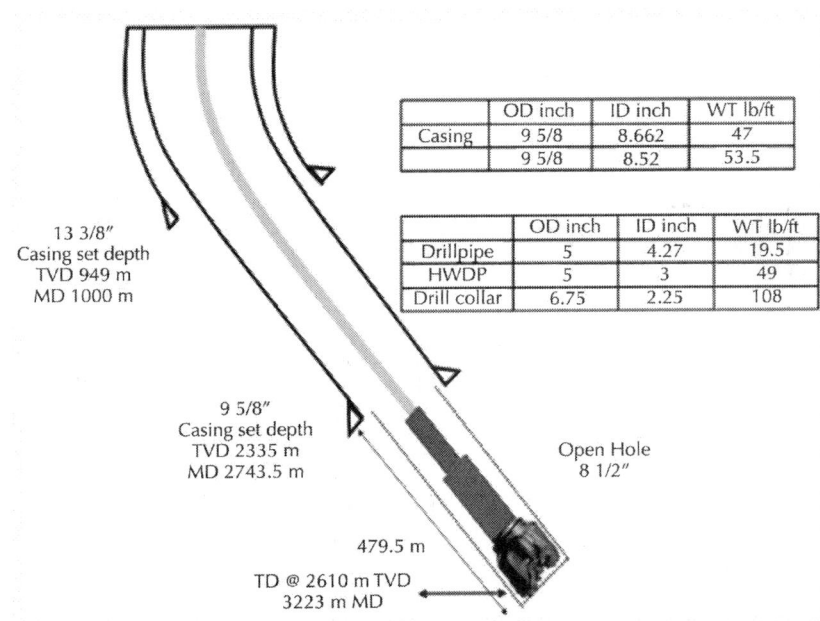

	OD inch	ID inch	WT lb/ft
Casing	9 5/8	8.662	47
	9 5/8	8.52	53.5

	OD inch	ID inch	WT lb/ft
Drillpipe	5	4.27	19.5
HWDP	5	3	49
Drill collar	6.75	2.25	108

13 3/8″
Casing set depth
TVD 949 m
MD 1000 m

9 5/8″
Casing set depth
TVD 2335 m
MD 2743.5 m

Open Hole
8 1/2″

479.5 m

TD @ 2610 m TVD
3223 m MD

Figure 11: Well schematic of Gulf of Suez oil field area.

Table 5 highlights the savings in total rig days and cost for conventional versus underbalanced drilling wells in Iran (Roving and Reynolds, 1994). It is clear that big savings in drilling cost was realized.

Table 5: Drilling time and cost savings for 8-1/2" hole section drilled underbalanced conditions

Well	Real cost		Clean cost (just drilling)	
	Days	K$	Days	K$
8-1/2" hole – conventional				
1	27	1171	27	1171
2	25.7	1146.3	24.4	1114
3	30.4	2125.3	21.6	1771.9
4	19.3	1360.1	17.6	1230.8
5	31.9	2215.7	16.7	1629.3
6	23.3	1058.5	22.4	1035
7	31.4	1385.1	23	1005.6
8	21.6	1241.5	17.8	989.9
9	20.7	899.1	17.2	667.4
10	34.1	1551.6	30.3	1300.1
Average	26.5	1415.4	21.8	1191.5
8-1/2" hole – underbalanced				
1	20.5	1652	14.8	1395.6
2	19	1458	13.7	1243.5
3	21.2	1998.6	16.5	1541.5
4	17.8	1193.6	15.7	728
5	12.9	597	12.2	553.9
Average	18.3	1379.8	14.6	1092.5

The cost savings ranged between $90,000 and $110,000 for 8-1/2 in. hole section and between $170,000 and $190,000 for the 6-1/2 in. hole size (Table 6). A total of approximately $1.4MM has been saved (drilling only) and about $1MM (overall), for the five wells drilled.

Table 6: Drilling time and cost savings for 6-1/2″ hole section drilled underbalanced conditions

Well	Total cost		Drilling cost	
	Days	K$	Days	K$
6-1/2″ hole – conventional				
1	9	886.6	9	886.6
2	11.8	591.8	11.8	591.8
3	20.7	1186.4	18.1	1082
4	29.6	1596.7	17.8	644.7
5	33.5	2074.1	20	1531.9
6	21.9	928.1	19.7	779.9
7	19.1	995.5	17.8	938.3
8	14.1	778.5	11.8	650.6
9	16.4	800.8	16.4	800.8
Average	19.6	1093.2	15.8	878.5
6-1/2″ hole – underbalanced				
1	7.4	507.8	6.6	471.9
2	24	1664.6	11.9	998.9
3	22.4	1804	17.2	1057.7
4	14.8	545.1	10.8	387.57
5	9.5	580.6	9	560.6
Average	15.6	920.4	11.1	695.3

PROPOSED UBD PROGRAM TO BE IMPLEMENTED IN EGYPTIAN FIELDS

Based on the experience and the problem faced discussed in the previous discussions, a proposed UBD program is given here-below.

Gulf of Suez Oil Field Area

The selected example includes drilling through the reservoir section, which consists of two production formations (Belayim and kareem formation from Miocene age). The reservoir and formation characteristics are given in Table 7 and Table 8.

Table 7: Gulf of Suez reservoir characteristics

Parameter	Belayim	Kareem
Pressure	1500 psi	1700 psi
Temperature	180 °F	190 °F
Gas–oil ratio (GOR)	15–17 SCF/STB	20 SCF/STB
Porosity (md)	18–20%	20–22%
Permeability	200 md	500 md
API^0 gravity of oil	20–23	20–30
H$_2$S concentration	No	No

Table 8: Gulf of Suez formation characteristics

Formation	Lithology	Top (m)	Thickness (m)	Pore pressure (psi)
Belayim				
Hammam Faraun	Shale-sand	2160	35	
Ferran	Shale-sand	2195	140	
Sidri	Mainly sand	2335	65	1500
Babaa	Anhydrite	2400	15	
Kareem	Limestone	2415	195	1700

The selected reservoir can be drilled by underbalanced drilling technique and the proposed UBD program is given in Table 9. Fig. 12 shows the operating window, multiphase fluid injection of Gulf of Suez oil field area.

Table 9: Underbalanced drilling design parameters for Gulf of Suez area

Rig modification	No essential modifications to be made on the rig to suite UBD operations • The substructure has to be high enough to allow Rotating Control Head (RCH) to be installed on top of the Hydril
Well plan	• As shown in Fig. 11
Drill string design	• Use a 5" DP and 5" HWDP on 6-3/4" DC
BHA	• The BHA consists of 6-1/2" mud motor and MWD to drill 8-1/2" hole • An 8-1/2" bit size of 3 × 13/32" nozzles
Drilling fluid selection	• The deviated section will be drilled using an oil bas mud and a membrane nitrogen generation circulating system
A-liquid phase	• Drilling fluid is native crude oil with density 7.6 ppg (0.91 S.G. or 20° API) • Liquid flow rates were selected to achieve a drawdown from the reservoir pressure
B-gas phase	• Nitrogen was selected as the injection gas • Nitrogen will be obtained from the surrounding air and generated onsite
Operating envelope	• A minimum drawdown at the bit of 100 psi is required to ensure adequate underbalanced conditions in the well • Using 300 gpm and more than 2400 scfm of Nitrogen will provide maximum 100 psi drawdown from the expected reservoir pressure, as shown in Fig. 12 • In case the real reservoir pressure will result below the expected value, then the liquid injection rate should be reduced increasing the risk for a hole cleaning issue
Hole cleaning	• Minimum annular liquid velocities in deviated holes of 210 ft/min when crude oil is used as the drilling fluid to ensure that the drilled cuttings are effectively removed from the wellbore • A wiper drilling trip will help clear the problem of hole cleaning

Motor performance	• The motor should be suitable for oil/nitrogen two-phase application • A maximum Equivalent Liquid Volume through the motor of 600 gpm was used as reference • A pressure loss of 800 psi between downhole motor and MWD was considered • The motor should not have a bypass valve on top of it
Production sensitivity	• As more reservoir fluids (oil and gas) introduced into the wellbore, the bottomhole circulating pressures (BHCP) will decrease • BHCP will therefore be controlled by increasing liquid injection and/or decreasing nitrogen injection, based on real-time BHCP data from the MWD tool • BHCP could also be controlled with surface backpressure • Choking will be necessary in stabilizing the circulating system during and after drill string connections
Data acquisition	• The software for the rig data acquisition has to be able to interface with the UBD equipment software
Completion	• The well can be completed with barefoot completion technique, or installing a slotted liner completions

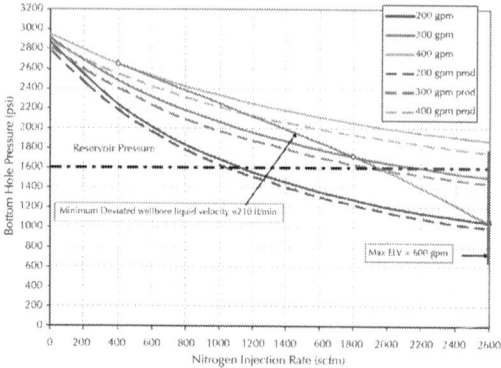

Figure 12: Operating window, multiphase fluid injection of Gulf of Suez oil field area.

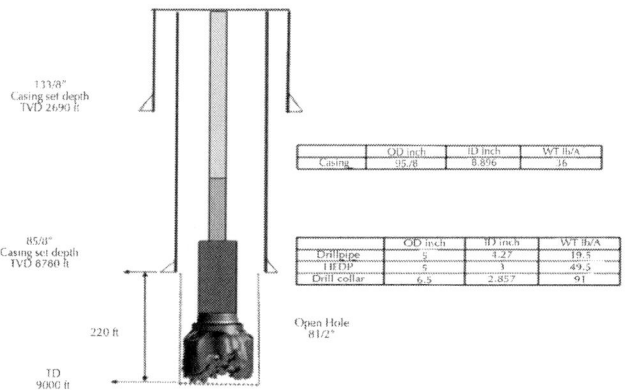

Figure 13: Well schematic of western desert oil field area.

Western Desert Oil Field Area

The selected example includes drilling through the reservoir section, which consists of Alam El Buieb formation of Cretaceous age. The lithology of this formation is sandstone with depleted reservoir pressure 1600 psi, reservoir temperature 219 °F, porosity 19%, permeability 200 md, GOR 95 SCF/STB, 41.7° API gravity of oil, and there is no H_2S concentration. The selected reservoir can be drilled by underbalanced drilling technique as given in Table 10. Fig. 14 shows the operating window, multiphase fluid injection of western desert oil field area.

Table 10: Underbalanced drilling design criteria for western desert area

Rig modification	• No essential modifications to be made on the rig to suite UBD operations • The substructure has to be high enough to allow Rotating Control Head (RCH) to be installed on top of the Hydril
Well plan	• As shown in Fig. 13
Drill string design	• Use 5″ DP, 5″ HWDP and 6.5″ DC

BHA	• •No downhole motor used • •An 8-1/2″ bit size of 3 × 13/32″ nozzles size
Drilling fluid selection	• Based on the pore pressure and formation depth, the reservoir formation is below the normal pressure regime • The subnormal pressure requires the use of a multiphase (liquid + gas) drilling fluid system in order to obtain on Underbalanced drilling condition
A-liquid phase	• Drilling fluid is native crude oil with density 6.84 ppg (0.82 S.G. or 41.7° API) • Liquid flow rates were selected to achieve a drawdown from the reservoir pressure
B-gas phase	• Nitrogen was selected as the injection gas
Operating envelope	• It is displayed as the area of the graph between the targets BHCP's, bound by the maximum motor throughput, the minimum annular liquid velocity, Fig. 11 • Using 300 gpm and more than 2200 scfm of Nitrogen will provide maximum 200 psi drawdown from the expected reservoir pressure
Hole cleaning	• Depends on several variables such as cutting size and shape; liquid properties; drill string rotation; liquid velocities; flow regime, etc. • Minimum vertical annular liquid velocities of 180 ft/min when crude oil is used as the drilling fluid to ensure that the drilled cuttings are effectively removed from the wellbore
Hydraulic modeling	• Using a multiphase hydraulic simulator, the required underbalanced drilling parameters could be evaluated in detail • Graphs can be created to incorporate the limiting factors of minimum annular liquid velocity required for hole cleaning and the desired BHCP range
Pressure while drilling	• When the maximum gas volume fraction (GVF) inside the drill pipe is bellow, 20% conventional mud pulse tools (MWD/LWD/PWD) can be used • Otherwise, electromagnetic transition tools have to be used in order to obtain downhole data real time
Data acquisition	• The software for the rig data acquisition has to be able to interface with the UBD equipment software

Completion	• The well can be completed with barefoot completion technique, or installing a slotted lined

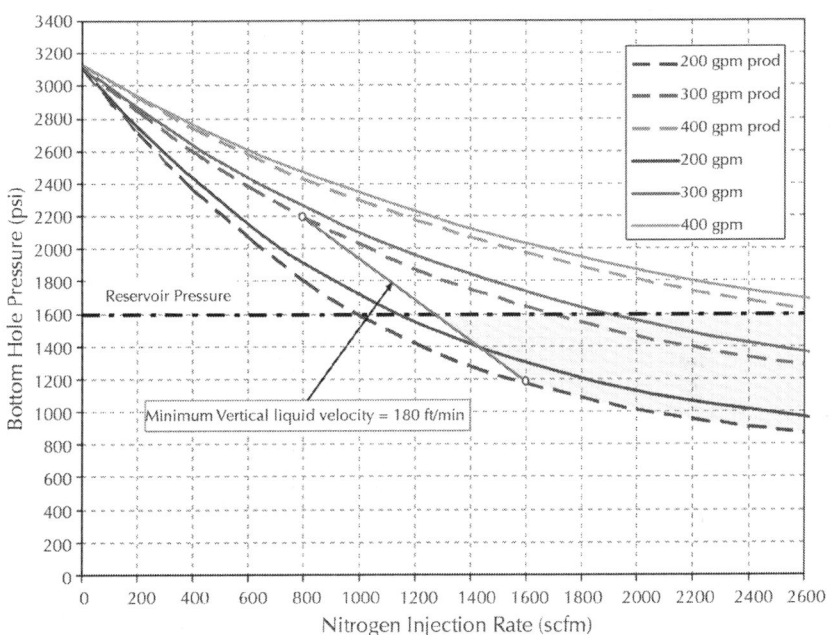

Figure 14: Operating window, multiphase fluid injection of western desert oil field area.

Nile Delta Oil Field Area

The selected example includes the reservoir section, which consists of one production formation (Qawasim from Miocene age). It has a sandstone lithology with reservoir pressure 3800 psi, reservoir temperature 185 °F, GOR 1100 SCF/STB, average porosity 25%, average permeability 400 md, gravity of oil 50° API, and there is no H_2S concentration.

The selected reservoir can be drilled by underbalanced drilling technique as given in Table 11. Fig. 16shows the operating window, multiphase fluid injection of nile delta oil field area.

Table 11: Proposed UBD program in Nile Delta area

Rig modification	• No essential modifications to be made on the rig to suite UBD operations • The substructure has to be high enough to allow Rotating Control Head (RCH) to be installed on top of the Hydril
Well plan	• As shown in Fig. 15
Drill string design	• Use a 5″ DP, 5″ HWDP and 6.5″ DC • An 8-1/2″ bit size of 3x13/32″ nozzles
BHA	• The BHA consists of 6-1/2″ PDM mud motor and MWD to drill 6″ hole • If MWD signal doesn't observed, use electromagnetic MWD tools
Drilling fluid selection	• Water based fluid (flow-drilling operation) • Drilling fluid is water with density 8.75 ppg (1.05 S.G.) • Liquid flow rates and surface choke backpressure were selected to achieve a drawdown from the reservoir pressure
Operating envelope	• It is recommended to pump at least 400 gpm of liquid phase to avoid any operational problem related with hole cleaning • The drawdown is 200 psi to prevent wellbore collapse
Motor performance	• A maximum equivalent liquid volume through the motor of 600 gpm was used as reference • A pressure loss of 800 psi between downhole motor and MWD was considered
Hole cleaning	• Minimum annular liquid velocities in deviated holes of 180 ft/min to ensure that the drilled cuttings are effectively removed from the wellbore • A wiper trip will help clear the hole cleaning problem
Tripping	• Some type of snubbing device can be used, or a downhole isolation valve can be installed • Balancing the well for trips seemed the simplest and least expensive method
Data acquisition	• The software for the rig data acquisition has to be able to interface with the UBD equipment software
Completion	• The well can be completed with barefoot completion technique, or installing a slotted lined

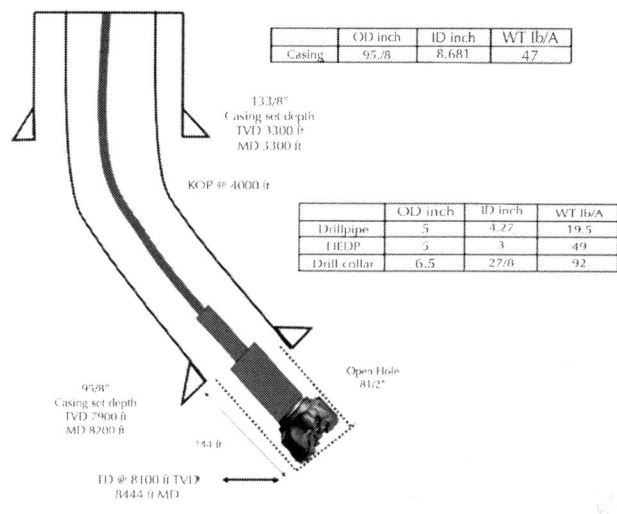

	OD inch	ID inch	WT lb/A
Casing	95./8	8.681	47

13.3/8"
Casing set depth
TVD 3300 ft
MD 3300 ft

KOP @ 4000 ft

	OD inch	ID inch	WT lb/A
Drillpipe	5	4.27	19.5
HEDP	5	3	49
Drill collar	6.5	27/8	92

95/8"
Casing set depth
TVD 7900 ft
MD 8200 ft

Open Hole
81/2"

TD @ 8100 ft TVD
8444 ft MD

Figure 15: Well schematic of Nile delta oil field area.

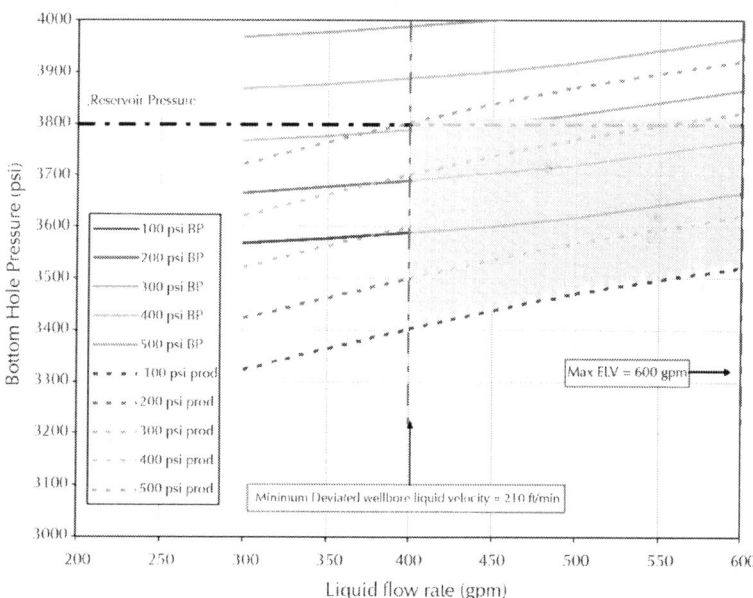

Figure 16: Operating window, flow-drilling operation for Nile delta oil field area.

CONCLUSIONS

Planned and applied correctly, underbalanced drilling technology can address problems of formation damage, lost circulation and poor penetration rates. The ability to investigate and characterize the reservoir while drilling is another important benefit of under balanced drilling. Based on the analysis of the real cases studied during the research, the following conclusions could be cited:

- Underbalanced drilling technique is a very useful technique especially when applied in reservoir section. It prevents formation damage, increases ROP, increases reservoir productivity and reduces the total cost of the well.

- Candidate screening is a rigorous and is a critical first step in the design of a successful underbalanced drilling operation. Although UBD has many advantages, it is not a magic solution for all fields or drilling problems. Poor screening and planning would result in an over-enthusiastic misapplication of the technology, and possibly failure.

- Many issues must be considered when designing an underbalanced drilling project including but certainly not limited to rock properties, reservoir pressure, borehole stability, drilling fluid type, injection method for gas assist, effect of compressible fluid on MWD, downhole motor requirements, bit type, corrosion, equipments availability, separation and fluid handling requirements especially when dealing with hydrocarbon drilling fluid, tripping procedures, data acquisition and completion procedures. Proper planning and design work, addressing these parameters, is essential to successfully conduct an underbalanced drilling project.

- UBD with stable foam through depleted reservoirs can be conducted safely and successfully in both vertical and horizontal wells. Drilling with foam has some appeal because foam has some attractive qualities and properties with respect to the very low hydrostatic densities, which can be generated with foam systems. Foam has good rheology and excellent cutting transport properties.

- Real time capture of production data while drilling should provide information about the reservoir not otherwise available.
- A proposed UBD program to be implemented in Egyptian fields is developed.

REFERENCES

1. Azeemddin, M. et al., 2006. Underbalanced Drilling Borehole Stability Evaluation and Implementation in Depleted Reservoirs, a Joaquin Field, Eastern Venezuela. IADC/SPE99165, February, 2006.

2. Bates, R.E., 1965. Field Results of Percussion Air Drilling. SPE 886, March, 1965.

3. Bennion, D.B., Thomasand, F.B., Bietz, R.F., 1998. Underbalanced Drilling: Praises and Perils. SPE Drilling and Completion, December, 1998.

4. Bentsen, N.W., Veny, J.N., 1976. Preformed Stable Foam Performance in Drilling and Evaluating Shallow Gas Wells in Alberta.

5. SPE 5712-PA, Formation Damage Conference held in Houston, October, 1976.

6. Black, A.D., Green, S.J., 1978. Laboratory simulation of deep well drilling. Pet. Eng., 40.

7. Bourgoyne, A.T., Young Jr., F.S., 1974a. A multiple regression approach to optimal drilling and abnormal pressure detection. SPEJ, 371.

8. Bourgoyne, A.T., Young Jr., F.S., 1974b. A multiple regression approach to optimal drilling and abnormal pressure detection. Trans. AIME, 257.

9. Boyun, Guo, Rajtar, J.M. 1995. Volume Requirements for Aerated Mud Drilling. SPE 26956-PA, Drilling and Completion, California Regional Meeting held in Ventura,

September, 1995.

10. Claytor, S.B., Manning, K.J., Schmalzried, D.L., 1991. Drilling a Medium-radius Horizontal Well with Aerated Drilling Fluid: A Case Study. Paper SPE 21988 presented at the 1991 SPE/IADC Drilling Conference, Amsterdam, March 11–14.

11. Cunningham, R.A., Eenink, J.G., 1959. Laboratory study of effect of overburden, formation, and mud column pressures on drilling rate of permeable formations. Trans. AIME 216, 9.

12. Dorenbos, Roelien, Ranalho, Jone, 2002. Underbalanced Drilling Primer. Shell International Exploration and Production B.V., June, 2002.

13. Eckel, J.R., 1957. Effect of pressure on rock drillability. Trans. AIME 213, 1.

14. Gamier, A.J., van Lingen, N.H., 1959. Phenomena affecting drilling rates at depth. Trans. AIME 216, 232.

15. George, E., Waston, Ralpha, A., 1956. Review of Air and Gas Drilling. SPE 703-G, Petroleum Branch Fall Meeting in Los Angeles, October, 1956.

16. Godwin, A., Lokpobiri, Ikoku, Chi U., 1986. Volumetric Requirements for Foam and Mist Drilling Operations. SPE 11723-PA, Petroleum Branch Office, California Regional Meeting held in Ventura, February, 1986.

17. Gordon, D. et al., 2005. Underbalanced Drilling with Casing Evolution in the south Texas Vicksburg. SPE Drilling and Completion, June, 2005.

18. Gray, Kenneth E., 1957. The Cutting Carrying Capacity of Air at Pressure above Atmospheric. SPE 874-G, October, 1957.

19. Hongren, G.U., Walton, J.C., Stein, D.A., 1999. Designing under- and near-balanced coiled-tubing drilling by use of computer simulations. SPE Dril. Comp. 14 (2).

20. Hooshmandkoochi, A., Zaferanich, M., Malekzadeh, A., 2007. First Application of Underbalanced Drilling in Fractured Carbonate Formations of Iranian Oil Fields Leads to Operational Success and Cost Saving. SPE 105536-MS, Middle

East Oil and Gas Conference held in Bahrain International Exhibition Center, Kingdom of Bahrailn, March, 2007.

21. International Association of Drilling Contractor, 2005. IADC Well Classification System for Underbalanced Operations and Managed Pressure Drilling <http://www.iadc.org/committees/underbalanced/>, March, 2005.

22. Kuru, E. et al., 1999. New Directions in Foam and Aerated Mud Research and Development. SPE 53963-MS, Latin American Caribbean Petroleum Engineering Conference held in Caracas, Venezuela, April, 1999.

23. Louison, R.F., Reese, R.T., Andrews, J.P., 1984. Case History: Underbalance Drilling the Midway and Navarro Formations Successfully in Hallettsville, TX. SPE13112, September, 1984.

24. Maclovio, Yanez M., 1996. PEP Region Norte and Valenzuela J. Marten, Tecominoacan 408: First Underbalance application in MEXECO. SPE 35320, March, 1996.

25. Maurer Engineering Manual, 1998. Underbalanced Drilling and Completion Manual, November, 1998.

26. Meng, Y. et al., 2005. Discussion of Foam Corrosion Inhibition in Air Foam Drilling. SPE 94469-MS, International Symbosium on Oil Field Corrosion held in Aberdeen, United Kingdom, May, 2005. Moore, C.L., Lafave, V.A., 1956. Air and Gas Drilling. SPE 494-G, February 1956.

27. Moore, D.D., Bencheikh, A., Chopty, J.R., 2004. Drilling Underbalanced in Hassi Messaud. SPE/IADC 91519, October, 2004.

28. Murray, A.S., Cunningham, R.A., 1955. Effect of mud column pressure on drilling rates. Trans. AIME 204, 196.

29. Nas, S., 2004. Leading Edge Advantage Ltd – Introduction to Underbalanced Drilling Manual, February, 2004.

30. Negra, A.F., Lage, A.C.V.M., Cunha, J.C., 1999. An Overview of Air/ Gas/Foam Drilling in Brazil. SPE 56865-PA, Drilling and Completion 14 (2), Drilling Conference held in Amsterdam, June, 1999.

31. Parra, J.G., Cells, E., Gennare, S., 2003. Wellbore Stability Simulations or Underbalanced Drilling Operations in Highly Depleted Reservoirs. SPE Drilling and Completion, June, 2003.

32. Qutob, H.H. et al., 2007. The Successful Application of Underbalanced Drilling Technology for Reservoir Evaluation and Drilling Performance Improvement in Kuwait. SPE 106680, June, 2007.

33. Qutob, Hani, Ferreira, Horacio, 2005. The SURE way to Underbalanced Drilling. SPE 93346, March, 2005.

34. Rankin, M.D., Friesenhahn, T.J., Price, W.R., 1989. Lightened Fluids Hydraulics and Inclined Bore Holes. Paper SPE 18670 presented at the 1989 SPE/IADC Drilling Conference, New Orleans, Feb. 28– March 3.

35. Roving, J.W., Reynolds, E., 1994. Underbalanced Drilling Through Oil Production Zones With Stable Foam in Oman. IADC/SPE 27525, February, 1994.

36. Salah El-Din, M.A., El-Katatney, S.M. (2009). Implementation of Underbalanced Drilling Technique in Egyptian Fields. M.Sc. Thesis, Cairo University, Egypt, 2009.

37. Sunthankar, A.A. et al., 2001. New Developments in Aerated Mud Hydraulics for Drilling in Inclined Wells. SPE67189, March, 2001.

38. Vidrine, D.J., Benit, E.J., 1968. Field verification of the effect of differential pressure on drilling rate. JPT, 676.

39. Weatherford Company, 2006. Operational Sequence in UBD (ROAD MAP). Weatherford Controlled Pressure Drilling and Testing Services.

40. Westermark, R.V., 1986. Drilling with a Parasite Aerating String in the Disturbed Belt, Gallatin County, Montana. IADC/SPE 14734, February, 1986.

41. Whiteley, Maxwel C., England, William P., 1986. Air Drilling Operation Improved by Percussion-Bit/Hammer-Tool Tandem. SPE Drilling Engineering, October, 1986.

42. Zhou, L. et al., 2005. Hydraulics of Drilling with Aerated Mud under Simulated Borehole Conditions. SPE/IADC 92484, February, 2005.

Forces Exerted on the Tool-electrode during Constant-Feed Glass Micro-drilling by Spark Assisted Chemical Engraving

Jana D. Abou Ziki and Rolf Wüthrich

Department of Mechanical and Industrial Engineering, Concordia University, Montreal, Quebec, Canada H3G 1M8

ABSTRACT

The forces exerted on the tool-electrode during Spark Assisted Chemical Engraving (SACE) constant velocity-feed glass micro-drilling are measured for different machining voltages, tool feed-rates and tool sizes. A diagram of the force regions in the hole-depth vs. tool feed-rate plane is constructed for different voltages and tool sizes. Two rate limiting steps for micro-drilling were identified. For low depths, the rate limiting step is the work-piece surface heating while for high depths it is the electrolyte flushing. Based on these findings, the tool feed-rate vs. hole-depth plane of the force regions was normalized using the time needed to heat the local glass surface and the tool radius. A correlation between the force occurrence and the current signal is identified where the current shifts upwards by a constant value when a force is exerted on the tool. This finding allows the usage of the current signal to detect the contact between the tool and the glass surface. The measurement and understanding of the forces exerted on the tool-electrode that this work brings is a first step towards the development of force feed-back algorithms for SACE machining.

INTRODUCTION

Spark Assisted Chemical Engraving (SACE), known as well in the literature under the names Electro Chemical Discharge Machining (ECDM) and Electro Chemical Spark Machining (ECSM), is one among a number of glass micro-machining technologies. This non-traditional technology, is based on electro-chemical discharge phenomena [1] and [2]. Machining occurs due to thermal assisted etching [3], [4], [5], [6], [7], [8], [9] and [10]. The heat provided by the electrochemical discharges allows the temperature to reach about 500–600 °C in the machining zone [11], [12], [13] and [14]. This accelerates etching of the work-piece by means of the OH radicals supplied from the electrolytic solution in the electrochemical cell based on the following reaction [6], [9] and [19]

$$2xNaOH + xSiO_2 \quad xNa_2 SiO_3 + xH_2O \tag{1}$$

During machining, the product of reaction (1) is evacuated out of the machining zone by the flowing electrolyte. As the temperature in the machining zone reaches the glass transition temperature, a glass layer of reduced viscosity forms below the tool. This layer will be referred to as "glass melt" in this text. The formation of such a layer was already reported in the case of 2D glass machining by SACE [15].

So far, several successful studies about minimization of geometrical errors of machined surfaces were reported in case of drilling and machining 2.5D structures with SACE [16], [17] and [18]. However, the process is still blind machining due to the lack of a suitable feedback signal to give sufficient information about the machining status. Further, like other micro-machining technologies such as micro Electrical Discharge Machining (μ-EDM) and micro Electro Chemical Machining (μ-ECM), SACE faces the problem of flushing the machining spot in case of micro-drilling. Contrary to μ-EDM and μ-ECM, which use the gap voltage or current signal to monitor the tool-work piece gap in order to optimize material removal [19],[20] and [21], the last could not yet be achieved in SACE machining. As a result, the most popular SACE drilling strategy is gravity-feed drilling. In addition to inadequate flushing of the machining spot, this approach suffers from tool bending, where both are serious problems especially when using micro-tools[22].

Although the current signal proved to be useful for extracting information about the machining process [23], the feasibility of using it as a feedback signal is still to be demonstrated. In this paper, an alternative signal to potentially develop feedback algorithms is explored: the force exerted on the tool during micro-drilling. No knowledge regarding this signal is available as of today. It is the first time that the forces are measured and analyzed to extract information related to the machining process. The possibility of using the force as a feedback signal to monitor the SACE machining process is discussed. In addition, it is shown that the current signal can be used to detect the tool work-piece contact. This opens new possibilities to use the current signal to develop feedback algorithms for SACE machining.

EXPERIMENTAL SETUP

The apparatus is composed of a machining head and a processing cell (Fig. 1). The machining head, mounted on the Z-axis of a

Cartesian robot, is composed of a flexible structure that can move freely in the z-direction parallel to the Z-axis, adding up a degree of freedom to the system. The machining head can be either used as a force sensor or as a profile-meter. The force sensor mode is obtained by using the zero-displacement force measurement principle. A PID controller maintains a fixed z position (measured with an optical sensor) by driving a voice coil actuator fixed on the flexible structure. The force needed to maintain a fixed z position is equal to the force exerted on the tool-electrode during machining. The PID controller was implemented using a dSPACE 1104 controller board. The force sensor has 100 ms response time to disturbance, an operating range of 0 to 5 N with 10 mN rms noise. In case the voice coil actuator is not driven (i.e. switched off), the machining head can be used as a profile-meter. The processing cell, in which the glass work-piece is fixed, is mounted on the XY stage to align the tool-electrode and the work-piece. An overflow system is used to maintain a fixed level of electrolyte above the work-piece (about 1 mm).

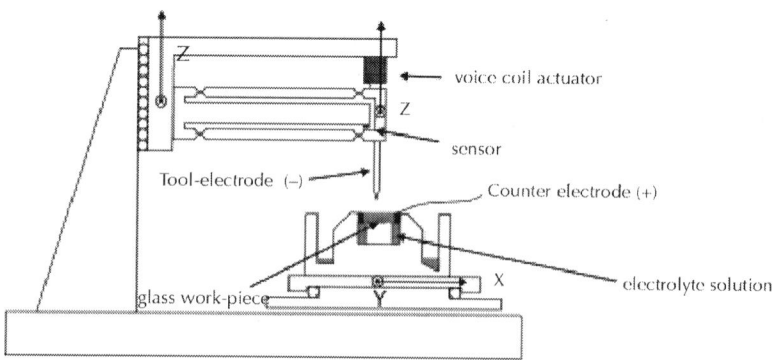

Figure 1: Schematic of SACE mechanical set-up which includes the machine head (holding the tool-electrode) mounted on a Z-stage and the cell (containing the glass slide) mounted on an XY stage. The machine head is composed of a flexible structure, to which the tool-electrode is attached, that can move freely in the z-direction. When used as a force sensor, a voice coil linear actuator is controlled to hold the structure at a fixed z position, measured by an optical sensor.

To study the forces exerted on the tool-electrode during micro-hole drilling, a set of experiments was conducted using different machining voltages (30 and 33 V), a range of tool feed-rates (1–80 µm/s) and two tool sizes (250 and 500 µm in diameter). The tool-electrodes are stainless steel (316 L) cylinders. The electrolyte is 30 wt% NaOH and the work-pieces are soda lime glass microscope slides (Bio Nuclear Diagnostics Inc.).

Prior to machining, the tool is positioned 50 µm above the glass surface (using the profile-meter function of the machining head). The machining voltage is switched on for 5 s in order to preheat the tool and the machining head is switched to the force sensor mode. Machining proceeds subsequently by moving the tool towards the glass at the specified feed-rate while recording the force signal. The holes machined were 250–400 µm deep. For each set of experiments 50 holes were produced to enhance the analysis reliability. Data was acquired using the ControlDesk software.

RESULTS AND DISCUSSION

Types of Forces

Based on the experimental results, three distinct force patterns are distinguished as depicted in Fig. 2. In the first configuration, the force signal increases linearly and saturates exponentially before dropping abruptly. In the second configuration, the force increases as in the first one, but then grows linearly further before dropping. In the third configuration, the forces increase linearly only. Based on the current understanding of the machining behavior, the following interpretation is proposed.

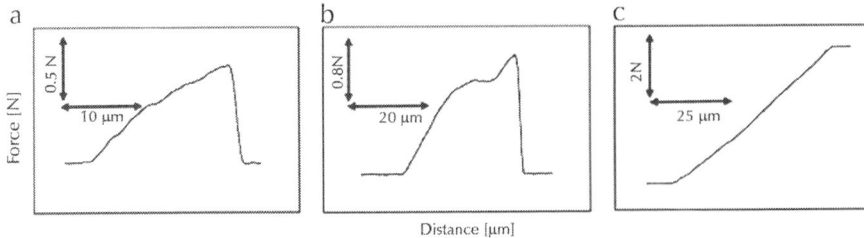

Figure 2: The three force patterns resulting from (a) on-going machining while the tool contacts the glass (recoverable force), (b) on-going machining with glass surface contact followed by negligible etching (recoverable force) and (c) negligible etching (unrecoverable force).

When no contact exists between the tool and the glass surface, no forces are present. As the tool touches the work-piece surface (Fig. 2a), a force appears. In the first configuration, the force increases and then grows at a reduced rate before recovering. A mechanism contributing to this force pattern is the on-going machining below the tool tip while the tool is pressing on the glass surface. These forces appear mainly near the work-piece surface in case of machining at high voltage (quicker surface heating). Note that in this configuration the forces lasted for a few microns before they disappeared abruptly due to the fast etching of the heated glass layer.

In the second configuration the force grows at a reduced rate, in a similar manner to the first one, depicting the on-going etching. However, afterwards the force increases again linearly, during a short time, before recovering. This increase is attributed to the low machining rate (dropping to almost zero) where for this case the slope of the force signal depicts the mechanical stiffness of the setup. This configuration was most often observed at depths higher than 100 μm where flushing is more difficult.

The third configuration (Fig. 2c) occurs either at the surface of the work-piece, when machining at both low voltage and high feed-rates, or at high depths. For low machining voltage the local machining zone is not hot enough to allow fast etching. Further, for high drilling depths (which vary depending on the tool feed-rate and the machining voltage; see Section 3.2), the material removal rate decreases significantly due to the lack of electrolyte supply inside the hole. In this case, the tool is always pushing on the glass surface and the forces are not recoverable. The measured slope is essentially determined by the stiffness of the mechanical setup.

Effect of Machining Voltage and Tool Feed-rate

To analyze the effect of heating on the machining forces, two machining voltages were selected (30 and 33 V) since they result in different discharge activity [1] causing different machining quality [24].

A broad range of tool feed-rates was investigated such that it allows obtaining minimal forces for up to 200 μm depth on one hand (low feed-rates) and the occurrence of large unrecoverable forces as soon as drilling starts on the other hand (high feed-rates).

Fig. 3 shows the recorded forces during successive drillings with various tool feed-rates for 30 V machining voltage. The origin of the Z-axis corresponds to the position where the experiment started (i.e. 50 μm above the initial glass surface when the system was at room temperature). Due to the high tool-electrode temperature, it expands typically by 10–15 μm [12]. This results in an early detection of forces after a tool downward motion by around 35–40 μm. In the following text these forces will be referred to as zero-depth forces.

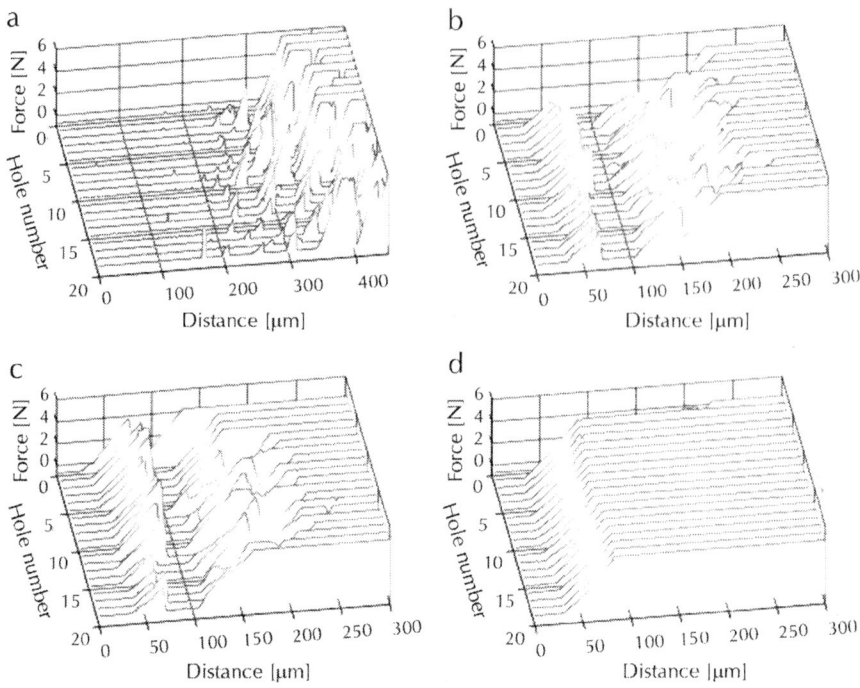

Figure 3: Recorded forces for different tool feed-rates: low (a: 1 μm/s), intermediate (b: 2 μm/s, c: 3 μm/s) and high (d: 5 μm/s). The machining voltage is 30 V and 500 μm tool-electrode was used. 20 holes were machined for each experimental set. Forces grow higher as the tool feed-rate increases where they saturated at 5 μm/s.

Three tool-feed ranges can be distinguished. For low feed-rates (Fig. 3a), no forces at zero-depth are observed and machining starts right away. At intermediate feed-rates, forces occur at zero-depth, but they disappear afterwards (Fig. 3b, c). Forces reappear only at higher depth. The region between the first and the second force appearance, during which no forces can be measured, is referred to as the *middle region*. At high feed-rates, forces are measured right from zero-depth and never recover in the present set-up (Fig. 3d).

Based on these measurements Fig. 4 is constructed which depicts schematically the regions where forces or no forces are observed in the hole-depth vs. tool feed-rate plane for different voltages and tool sizes. Fig. 4a depicts the force regions for 500 μm tool while applying both 30 and 33 V while Fig. 4b shows the force regions for a 250 μm tool and a 30 V machining voltage.

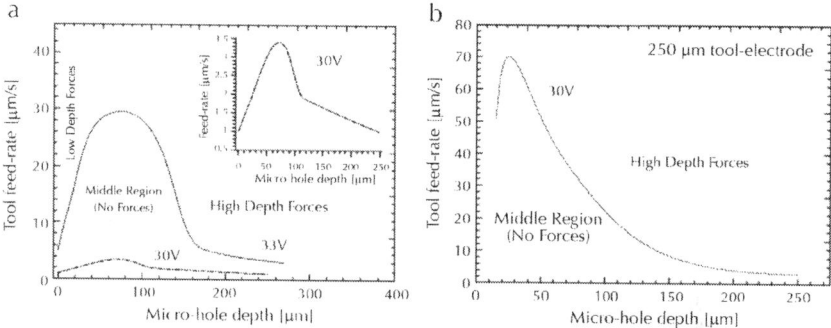

Figure 4: Schematic representation of the regions where forces or no forces are observed in the hole-depth vs. tool feed-rate plane while using 500 μm tool (a) and 250 μm tool (b). For (a), 30 and 33 V were applied where for 33 V higher range of tool feed-rates can be used. As shown in (b) for smaller tool size the allowable tool feed-rate range extends as well.

As shown in Fig. 4a, the low feed-rates correspond to feed-rates below 1 μm/s in the case of 30 V and below 5 μm/s for higher voltage, i.e. 33 V. For both voltages, forces start appearing at around 190 μm, recover quickly and reappear at about 250 μm depth. Forces at 190 μm depth are distributed between the three configurations (two and three dominating) while those at 250 μm belong mainly to the third type.

The intermediate tool feed-rates correspond to 1–4 μm/s at low voltage and to 5–30 μm/s at high voltage. Forces appear right at the initial work-piece surface but disappear rapidly. They reappear at around 60 μm depth for 30 V and at about 120–150 μm depth for 33 V. For the lower feed-rates, the zero-depth forces are distributed among the three configurations while they are reduced to forces in the second and third configurations with increasing feed-rates. The middle region decreases in width as the tool feed-rate increases and eventually vanishes for high tool feed-rates.

The tool feed-rate limit, above which only forces at zero-depth are observed (all belonging to the third configuration), is 5 μm/s for 30 V and increases to 30 μm/s for 33 V.

Note that a higher voltage extends the intermediate tool feed-rate range and it slightly widens the middle region. Similar to gravity-feed drilling [24], the holes were often deformed due to tool bending (Fig. 5a) and jagged contours appeared on the hole entrance at 33 V. Few holes with circular entrance (Fig. 5b) could be examined.

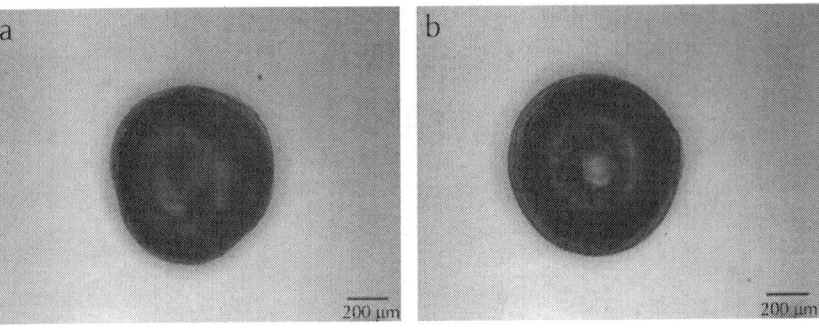

Figure 5: An example of the holes machined during constant velocity-feed machining at 33 V. Holes are often deformed due to tool bending (a) and jagged contours appeared on the hole entrance (similar to gravity-feed drilling). Holes with circular entrance (b) are less frequent.

Effect of Tool Size

To investigate the effect of the tool size on the forces, a tool of 250 μm diameter instead of 500 μm was used. For these experiments a 30 V machining voltage and a range of tool feed-rates were selected (Fig. 4b). Forces behave in a similar way as in the case of 500 μm tools. However, for smaller tools much higher feed-rates could be used.

No forces are observed at zero-depth for tool feed-rates up to 50 μm/s, compared to 1 μm/s in the case of the larger tool-electrode. Forces start appearing at around 70–190 μm for speeds between

30 and 3 µm/s respectively (Fig. 4b). For intermediate tool feed-rates (50–70 µm/s), forces appear right at the initial work-piece surface and reappear at around 50 µm depth. At high tool feed-rates (70 µm/s and higher), only forces at zero-depth which are unrecoverable could be observed as in the case of the 500 µm diameter tool-electrode.

Forces at Low and High Depths

In order to be able to machine, two conditions have to be satisfied. On one side the local temperature has to be high enough such that etching proceeds at notable rates. On the other side, electrolyte has to be available in the machining zone in order to provide the needed OH radicals for the glass etching. The etched material has to be removed from the machining spot afterwards. This is in accordance with the results found by Han et al. [25] where the surface roughness, which is reduced for appropriate heating and flushing conditions, follows an inverse volcano behavior in function of the tool feed-rate. This shows the importance of balancing both the hole-flushing (low feed-rates) and the surface heating (high feed-rates). As known from gravity-feed drilling experiments, different material removal rates will result depending on the micro-hole depth [22]. Based on these remarks, the described force behavior (3.1, 3.2 and 3.3) is discussed below.

Low Depth

For low machining depths (up to the beginning of the middle region), where the flushing of electrolyte is not the limiting step, machining occurs in the discharge regime as described in gravity-feed drilling [22]. The insufficient heating of the machining zone is the primary cause of the appearance of forces. Once the machining zone heats up, drilling can proceed at high material removal rates as known from gravity-feed drilling (about 50 µm/s for 30 V and 80 µm/s for 33 V while using a 400 µm tool-electrode [1]).

The typical time t_0 needed to heat up the glass surface in order to start machining, is given by [1] and [19]:

$$t_0 = \frac{\kappa^2 r^2}{4\pi a (\kappa - 1)^2}$$

(2)

for a tool-electrode of radius r and a work-piece of thermal diffusivity a. The number κ (normalized heat power) is the ratio between the heat P_0 transferred to the work-piece by the SACE process and the minimal heat P_{min} needed to machine. The accurate estimation of t_0 is difficult due to the sensitivity of t_0 to κ in Eq. (2). It is proposed to estimate P_0 by

$$P_0 = (U - U_d)I$$

(3)

with U the machining voltage, U_d the water decomposition potential (around 1.7 V) and I the mean current, which is typically about 0.01 A according to [12]. P_{min} can be estimated by [1] and [11]:

$$P_{min} = \lambda \Delta T \pi r$$

(4)

with λ the thermal conductivity of the work-piece and ΔT typically about 500–600 °C for glass machining[11], [12], [13] and [14].

In summary, the time needed to heat up the glass surface to allow machining is reduced by using a higher machining voltage or a thinner tool-electrode. Based on these remarks, the behavior observed at low-depth can be explained as follows:

For low tool feed-rates (as identified in Section 3.2), sufficient time is available for the machining zone to become hot enough to achieve notable glass etching before the tool-electrode touches the work-piece. No forces can be measured until reaching higher depths at which the electrolyte available is insufficient to allow efficient etching. The local temperature is still high, which heats up the glass surface layer. Fig. 6shows a typical example of the cooled melt formed when using a 500 μm tool-electrode (voltage and feed-rate being 30 V and 3 μm/s).

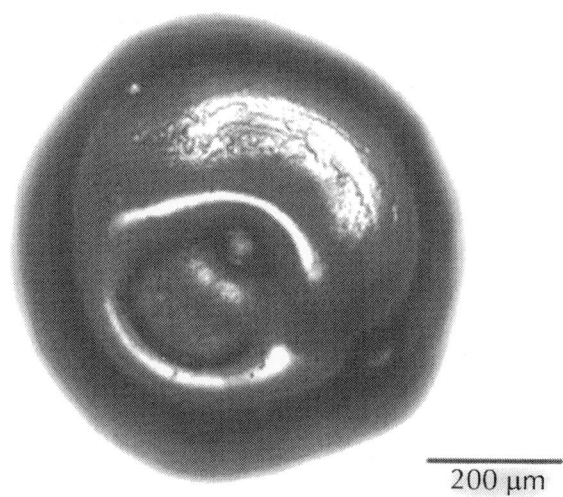

200 µm

Figure 6: A typical example of the cooled melt when using a 500 µm tool-electrode, 30 V and of 3 µm/s tool feed-rate.

For the intermediate tool feed-rates, less time is available to heat the work-piece which results in an earlier mechanical contact between the tool-electrode and the glass surface. Hence, zero-depth forces occur. As the surface is still "cold", these forces grow linearly (especially for higher tool-feed rates). Once the work-piece surface is heated enough, etching starts and is sustained as long as electrolyte is available in the machining zone (the extension of the middle region). This leads to a linear growth of the line separating the region of zero-depth forces and the middle region in the hole-depth, z, vs. tool feed-rate, F, plot (Fig. 4). The slope of this line can be calculated as the inverse of the time needed to heat the glass surface $t0$ given by Eq. (2)

$$F = \frac{1}{t_0}z$$

(5)

Based on the slope of these separation lines (Fig. 4), the time needed to heat up the work-piece is about 16 s for 30 V and 1.6 s for 33 V (for the 500 µm tool-electrode) and about 0.35 s for 30 V (for the 250 μm tool-electrode). These values agree well with the one predicted by Eq. (2) if using an average current of 0.01 A for the 500 μm tool-electrode and 0.007 A for the 250 μm tool-electrode. These current values are in agreement with the one needed to estimate the thermal expansion of the tool-electrode [12]. For high tool-feed rates the glass can hardly be heated locally. This is due to the increase in the force pushing on the work-piece which probably prevents proper gas film formation at the tool tip and therefore limits the discharges activity at this location. The altered formation of the gas film can also be seen through the current signal as will be further discussed in Section 3.6. Etching is inefficient since it can only happen from the side, resulting in very low material removal rates. Fig. 7 shows an example of a hole machined at 5 μm/s tool feed-rate and 30 V machining voltage. A bump 80 μm deeper than the hole entrance is formed in the middle of the hole which confirms previous findings while machining on ceramics [26] and [27]. For these conditions the hole is around 100 μm deep after 50 s machining time.

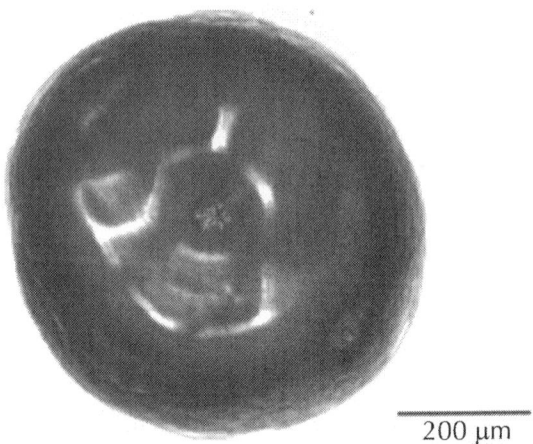

200 μm

Figure 7: An example of a hole machined at 30 V and 5 μm/s tool feed-rate. A bump 80 μm deeper than the hole entrance is formed in the middle of the hole. The hole is around 100 μm deep after 50 s machining time.

High Depth

For higher machining depths (beyond the middle region) flushing of the hole with electrolyte becomes the limiting step and the lack of OH radicals to conduct the etching is the main reason for the force appearance. Depending on the machining voltage and the tool feed-rate, the depth at which high forces occur varies. This can be seen by the variation in the width of the middle region prior to the appearance of the high depth forces. The line separating the middle region and the high force region represents the depth limit at which machining enters the hydrodynamic regime where forces are high and the machining rate is low.

As the tool feed-rate increases, the width of the middle region decreases since the material removal rate is expected to be lower in this case due to the increasing difficulty of flushing the machining zone and evacuation of the machined material. For low feed-rates, the high-depth forces following the middle region are composed of the three types. As the feed-rate increases, the forces are reduced to the third type only. This is attributed to the fact that for low feed-rates a melt can form in the middle region whereas for higher feed-rates the tool is always pushing on the "cold" bulk glass. At very high depths, the insufficient electrolyte available inside the hole significantly reduces the glass etching rate. For higher voltage, the middle region is wider due to the higher tool-electrode temperature which enhances the etching.

Dimensionless Description of Forces in the Tool Feed-rate vs. Hole-depth Plane

Based on conclusions from 3.4.1 and 3.4.2, it is proposed to write Eq. (5) in the following dimensionless form:

$$\frac{F}{r/t_0} = \frac{z_0}{r} \tag{6}$$

Eq. (6) suggests that the force characteristics in the tool feed-rate vs. hole-depth plane can be described in dimensionless form by normalizing the tool feed-rate by r/t_0 and the drilling depth by r. Applying this normalization to Fig. 4 results in the normalized plot of the force regions in the tool feed-rate vs. hole-depth plane as shown in Fig. 8.

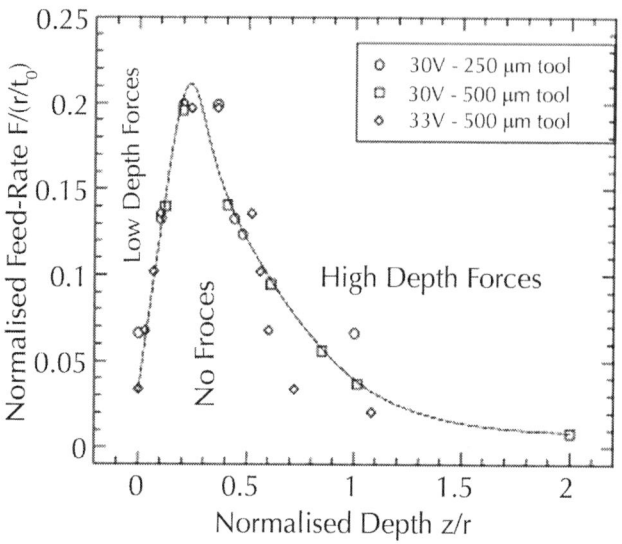

Figure 8: Plot of the regions where forces or no forces are observed in the normalized hole-depth vs. normalized tool feed-rate plane. Normalization is done according to Eq. (6).

From Fig. 8 it can be seen that the maximal tool feed-rate F_{max} below which a middle region exists is given by

$$F_{max} = 0.2 \frac{r}{t_0}$$

(7)

In gravity-feed drilling it was determined in the case of a 400 μm diameter tool-electrode that the hydrodynamic regime (i.e. where electrolyte flushing becomes the rate limiting step) starts after a drilling depth δ of around 70 μm (i.e. 0.35 in normalized depth) [11]. The distance δ is expected to be that at which the high-depth forces start to appear. This value is similar to the one resulting from Fig. 8 (i.e. z_{max}/r=0.2 forF_{max}), which is a good indication that indeed the origin of the high depth forces is the insufficient flushing of the machining spot.

Rate Limiting Steps during SACE Micro-drilling

In further support to the argument regarding the rate limiting steps while drilling, the following experiments were carried out. The tool (500 μm diameter) is moved downwards with a constant feed-rate while the force is measured. When the force level exceeds a predefined threshold, the tool is moved upwards by a certain distance. The experiments were conducted while varying the upward displacement (5–30 μm) and the tool feed-rate (3–30 μm/s). For each set of experiments 10 holes, 400 μm in depth, were machined.

Fig. 9a shows an example of the evolution of drilling depth in function of time for different tool feed-rates (3–30 μm/s), 33 V machining voltage, an upward motion of 10 μm and a force threshold of 850 mN.

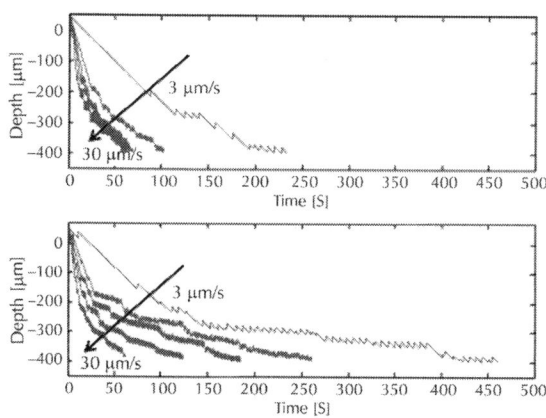

Figure 9: Drilling depth in function of time for different tool feed-rates (3–30 μm/s), 500 μm tool, 33 V. For all feed-rates, when the force exceeds a pre-set threshold the tool is moved upwards by 10 μm. top Corresponds

to 850 mN force threshold and bottom corresponds to 230 mN force threshold. The drilling time is reduced when increasing the force threshold and the tool-feed rate.

Results show that the chosen upward motion significantly affects the machining time. For all tool feed-rates, the drilling time, measured at 300 μm depth of the average drilling evolution, was minimal for 25 μm upward motion (Fig. 10). This behavior is attributed to the trade-off between efficient flushing (high gap) and efficient heating of the machining zone (low gap). A similar effect is observed in gravity-feed drilling in conjunction with tool-electrode rotation, tool vibration and when using flat-sided tool-electrode with pulsed voltage where all these effects enhance the flushing of the hole on the cost of efficient heating [17], [18],[28] and [29]. These results are also in agreement with the conclusions obtained when using different pulsed frequencies and duty ratios during SACE micro-drilling where it was demonstrated that for higher pulse frequency and lower duty ratio the thermal damage is reduced [16]. This shows that a key factor during SACE machining is the balanced heating and flushing of the hole.

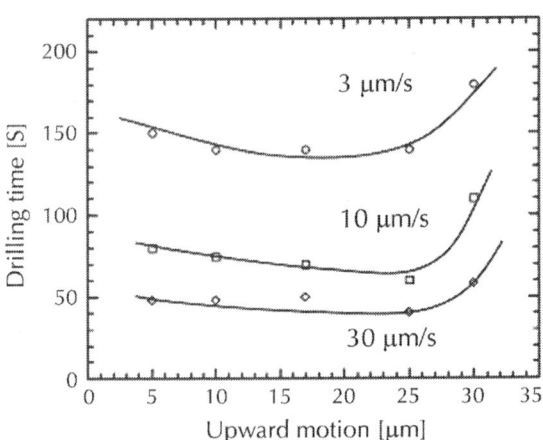

Figure 10: Drilling time measured at 300 μm depth of the average drilling evolution in function of tool upward motion for different tool feed-rates. For all tool feed-rates, the drilling time was minimal for 25 μm upward

motion which is attributed to a trade-off between efficient flushing and efficient heating of the machining zone.

To further support this hypothesis, experiments were repeated while reducing the force threshold from 850 mN to 230 mN. In this case, the local glass surface will be less heated due to the poorer contact between the tool and the surface. The resulting drilling time increases, which was most apparent for low tool feed-rates. For example, for 3 μm/s feed-rate and 10 μm upward motion, the drilling time is typically 450 s for the low force threshold compared to around 240 s for higher threshold (Fig. 9).

To confirm the importance of work-piece heating, a waiting time is added to heat the surface once the force threshold is exceeded. To ensure that the occurrence of forces right from the beginning of machining, the tool feed-rate was chosen to be 30 μm/s, which is the limiting feed-rate in case of 33 V machining voltage. For depths lower than around 300 μm, when applying 0.5 s waiting time the frequency of the forces is similar with and without flushing (Fig. 11a, b). For higher waiting time (5 s) and in the absence of flushing, the frequency of the forces is reduced (Fig. 11b, c). Thus, the rate limiting step is the localized heating of the glass surface for low depths. Note that the large waiting time needed to heat up the local glass surface indicates that the heat transfer between the tool and the glass surface is poor. This confirms the remarks stated in Section 3.2.

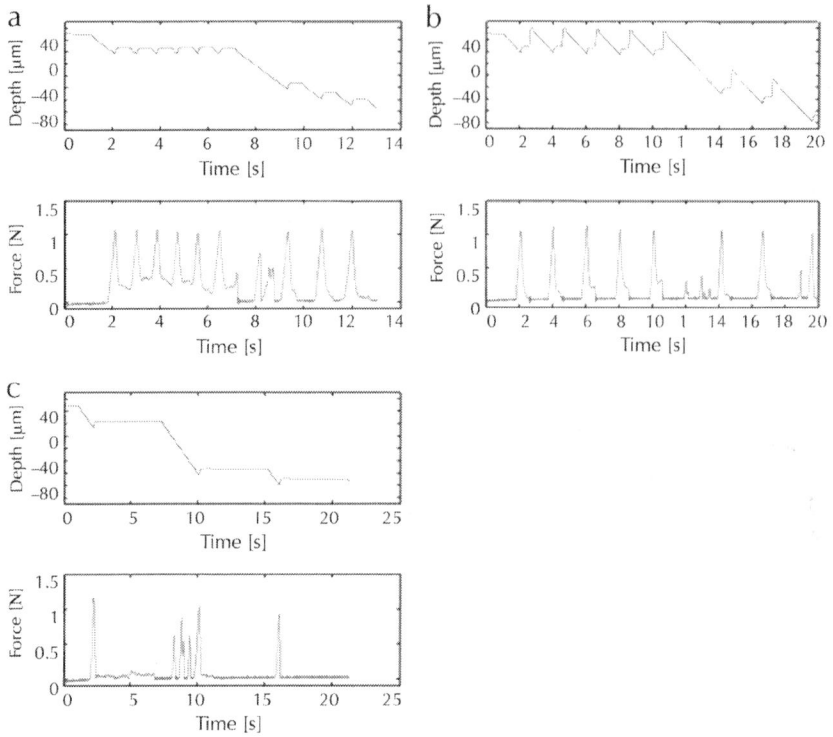

Figure 11: Drilling depth and force frequency in function of time (for depths lower than 300 µm) for 33 V, 30 µm/s and 0.5 s waiting time, a: with flushing; b: without flushing; c: without flushing and while increasing the waiting time to 5 s. No difference existed between cases (a) and (b), however for (c) drilling time was reduced.

For depths higher than around 300 µm, when combining heating and flushing, the machining rate was not enhanced. When only flushing is added, the material removal rate was rapid, similar to that at lower depths. The result was applicable for the complete tool feed-rate range. This indicates that for depths higher than 300 µm the rate limiting step is the electrolyte flow within the microhole.

Based on the mentioned outcomes, it can be concluded that depending on the hole depth the importance of the electrolyte flushing of the hole and the localized heating due to the discharges varies. Note that the significance of hole-flushing was already known for gravity-feed drilling experiments. The analysis of the forces showed further that the heat transfer to the work-piece is an important factor as well. For these experiments, most of the holes looked like the example shown in Fig. 5b. As expected, the tool bending was greatly reduced when limiting the force threshold. Similar to gravity-feed drilling, jagged contours were formed on the hole›s entrance [24].

Current Signal

A correlation between the current signal and the occurrence of forces during drilling could be observed.Fig. 12 shows the force and the corresponding current signal in function of time for 33 V machining voltage and 15 µm/s tool feed-rate. When a force is exerted on the tool-electrode, the current is shifted upwards by about 20 mA independently of the force amplitude. As soon as the force threshold (850 mN) is exceeded, the tool is moved upwards where the shift disappears in this case. The discharge activity remains unaltered.

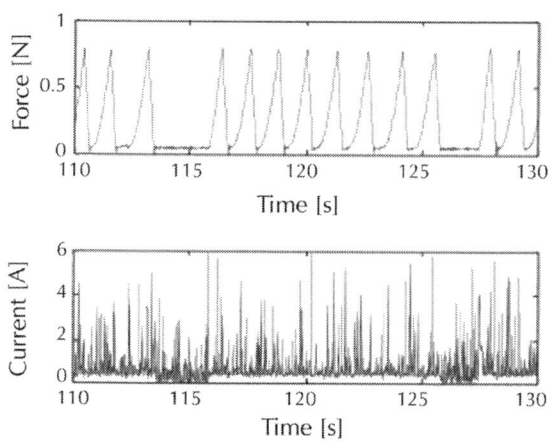

Figure 12: An example of the force signal and the corresponding current signal in function of time for 33 V and 15 µm/s tool feed-rate. When a

force is exerted on the tool-electrode, the current is shifted upwards by around 20 mA independently of the force amplitude. As the force is removed, the shift in the current signal disappears.

It is hypothesized that as the tool touches the work-piece, the gas film cannot form properly at the tool tip and is pushed upwards. The discharge activity at the tip is reduced and the discharges are generated on the side of the tool. A possible mechanism that explains the upward shift in the current signal is proposed. As the gas film is pushed upwards, a strand of electrolyte connects the upper part of the tool to the bulk electrolyte in the cell. This creates an additional path for the current signal. The observed correlation between the forces and the current opens up a new possibility in using the current signal as a way to detect the contact between the tool and the work-piece similar to the case of EDM [19] and [30]. As a result, some drilling algorithms based on contact detection between the tool and the work-piece during EDM or ECM may be adapted for SACE machining.

CONCLUSIONS

In this paper, the forces exerted on the tool-electrode during SACE micro-drilling were investigated for different machining voltages (30 and 33 V), tool feed-rates (1–80 μm/s) and tool sizes (250 and 500 μm in diameter). Three force patterns, determined by the weighted contribution of the etching progress vs. the tool mechanical contact with the glass surface, could be distinguished. For each machining voltage, three tool feed-rate ranges were identified based on the depth at which forces appear and their ability to recover. For high machining voltage (33 V) and reduced tool size (250 μm diameter), the allowable tool feed-rate range increases as the heating is enhanced in this case. The curves classifying the force regions in terms of the hole-depth and the tool feed-rate, for each machining voltage and tool size, are superposed on one curve by normalizing the tool feed-rate and hole-depth. Forces are categorized into high- and low-depth forces based on the force measurement. For low depths, the local glass heating is the rate limiting step. However, for high depths the hole-flushing is the rate limiting step.

A shift in the current is observed at the moment the tool contacts the surface. This can be explained by the additional current path created at the upper part of the tool as the gas film is pushed upwards. Hence, less discharges occur at the tip resulting in poor heat transfer from the tool-electrode to the work-piece. This finding allows using the current signal to detect the contact between the tool and the work-piece. The current understanding of the machining forces results in enhancing the knowledge of SACE machining process, therefore openning the possibility to enhance machining performance by developing force-feedback algorithms for SACE micro-drilling.

ACKNOWLEDGMENTS

This work was supported by the Natural Sciences and Engineering Research Council of Canada (NSERC). J.D.A.Z. would like to thank the Ministère de l'Education, du Loisir et du Sport du Quebec (MELS) for the bourse d'excellence pour étudiants étrangers (V1) and Posalux SA for the Posalux Excellence Scholarship.

REFERENCES

1.   R. Wüthrich Micromachining Using Electrochemical Discharge Phenomenon: Fundamentals and Applications of Spark Assisted Chemical Engraving William Andrew, Norwich (2009)

2.   R. Wüthrich, V. Fascio Machining of non-conductive materials using electrochemical discharge phenomenon—an overview International Journal of Machine Tools and Manufacture, 45 (2005), pp. 1095–1108

3.   C.T. Yang, S.S. Ho, B.H. Yan Micro hole machining of borosilicate glass through electrochemical discharge

machining (ECDM) Key Engineering Materials, 196 (2001), pp. 149–166

4.   I. Basak, A. Ghosh Mechanism of material removal in electrochemical discharge machining: a theoretical model and experimental verification Journal of Materials Processing Technology, 71 (1997), pp. 350–359

5.   V.K. Jain, P.M. Dixit, P.M. Pandey On the analysis of the electrochemical spark machining process International Journal of Machine Tools and Manufacture, 39 (1999), pp. 165–186

6.   V. Fascio, R. Wüthrich, H. Bleuler Spark assisted chemical engraving in the light of electrochemistry Electrochimica Acta, 49 (2004), pp. 3997–4003

7.   H. Tokura, I. Kondoh, M. Yoshikswa Ceramic material processing by electrical discharge in electrolyte Journal of Materials Science, 24 (1989), pp. 991–998

8.   C. Tsutsumi, K. Okano, T. Suto High quality machining of ceramics Journal of Materials Processing Technology, 37 (1993), pp. 639–654

9.   V. Fascio, R. Wüthrich, D. Viquerat, H. Langen, 3D microstructuring of glass using electrochemical discharge machining (ECDM), in: Proceedings of International Symposium on Micromechatronics and Human Science, 1999 pp. 179–183.

10.   V. Fascio, Etude de la Microstructuration du Verre par étincelage Assisté par Attaque Chimique: Une Approche Electrochimique (Ph.D. thesis), Swiss Federal Institute of Technology, EPF Lausanne, 2002.

11.   M. Jalali, P. Maillard, R. Wüthrich, Toward a better understanding of glass gravity-feed micro-hole drilling with electrochemical discharges, Journal of Micromechanics and Microengineering 19, 2009, 45001–45008.

12.   J.D. Abou Ziki, R. Wuthrich, Tool wear and tool thermal expansion during micro-machining by spark assisted chemical

engraving, International Journal of Advanced Manufacturing Technology 61, 2012, 481–486.

13. H.H. Kellogg, Anode effect in aqueous electrolysis, Journal of The Electrochemical Society 97 (1950) 133–142.

14. V. Reghuram, Electrical and Spectroscopic Investigations in Electrochemical Discharge Machining (Ph.D. dissertation), Indian Institute of Technology, Madras and Indian Institute of Technology, Kanpur, 1994.

15. J.D. Abou Ziki, T.F. Didar, R. Wüthrich, Micro-texturing channel surfaces on glass with spark assisted chemical engraving, International Journal of Machine Tools and Manufacture 57, 2012, 66–72.

16. D.J. Kim, Y. Ahn, S.H. Lee, Y.K. Kim Voltage pulse frequency and duty ratio effects in an electrochemical discharge microdrilling process of Pyrex glass International Journal of Machine Tools and Manufacture, 46 (2006), pp. 1064–1067

17. E.S. Lee, D. Howard, E. Liang, S.D. Collins, R.L. Smith Removable tubing interconnects for glass-based micro-fluidic systems made using ECDM Journal of Micromechanics and Microengineering, 14 (2004), pp. 535–541

18. N. Gautam, V.K. Jain Experimental investigations into ECSD process using various tool kinematics International Journal of Machine Tools and Manufacture, 38 (1998), pp. 15–27

19. Y.F. Luo The dependence of interspace discharge transitivity upon the gap debris in precision electrodischarge machining Journal of Materials Processing Technology, 68 (1997), pp. 121–131

20. K.P. Somashekhar, N. Ramachandran, J. Mathew Optimization of material removal rate in micro-EDM using artificial neural network and genetic algorithms Materials and Manufacturing Processes, 25 (2010), pp. 467–475

21. L. Yong, Z. Yunfei, Y. Guang, P. Liangqiang Localized electrochemical micromachining with gap control Sensors and Actuators A: Physical, 108 (2003), pp. 144–148

22. R. Wüthrich, U. Spaelter, Y. Wu, H. Bleuler A systematic characterisation method for gravity feed micro-hole drilling in glass with Spark Assisted Chemical Engraving (SACE) Journal of Micromechanics and Microengineering, 16 (2006), pp. 1891–1896

23. R. Wüthrich, U. Spaelter, H. Bleuler The current signal in spark-assisted chemical engraving (SACE): what does it tell us? Journal of Micromechanics and Microengineering, 16 (2006), pp. 779–785

24. P. Maillard, B. Despont, H. Bleuler, R. Wüthrich Geometrical characterization of micro holes drilled in glass by gravity-feed with spark assisted chemical engraving (SACE) Journal of Micromechanics and Microengineering, 17 (2007), pp. 1343–1349

25. M.S. Han, B.K. Min, S.J. Lee Improvement of surface integrity of electro-chemical discharge machining process using powder-mixed electrolyte Journal of Materials Processing Technology, 19 (2007), pp. 224–227

26. S.K. Chak, P.V. Rao Trepanning of Al2O3 by electro-chemical discharge machining (ECDM) process using abrasive electrode with pulsed DC supply International Journal of Machine Tools and Manufacture, 47 (2007), pp. 2061–2070

27. B.R. Sarkar, B. Doloi, B. Bhattacharyya Parametric analysis on electrochemical discharge machining of silicon nitride ceramics International Journal of Advanced Manufacturing Technology, 28 (2006), pp. 873–881

28. R. Wüthrich, B. Despont, P. Maillard, H. Bleuler Improving the material removal rate in spark-assisted chemical engraving (SACE) gravity-feed micro-hole drilling by tool vibration Journal of Micromechanics and Microengineering, 16 (2006), pp. 28–31

29. Z.P. Zheng, H.C. Su, F.Y. Huang, B.H. Yan The tool geometrical shape and pulse-off time of pulse voltage effects in a Pyrex glass electrochemical discharge microdrilling process Journal of Micromechanics and Microengineering, 17 (2007), pp. 265–272

30. H.D. Kauffman, M.F. Davis, EDM Process Method and Apparatus for Controlling the Flow Rate of Dielectric as a Function of Gap Impedance, U.S. Patent 3 699 303, 1972.

The Effect of Drilling Fluids and Crude Oil on Some Chemical Characteristics of Soil and Crops

Ivica Kisic[a], Sanja Mesic[b], Ferdo Basic[a], Vladislav Brkic[b], Milan Mesic[a], Goran Durn[c], Zeljka Zgorelec[a], and Lidija Bertovic[b]

[a]University of Zagreb, Faculty of Agriculture, Svetosimunska cesta 25, 10 000 Zagreb, Croatia

[b]INA-NAFTAPLIN, Subiceva 29, 10 000 Zagreb, Croatia

[c]University of Zagreb, Faculty of Mining, Geology and Petroleum Engineering, Pierottijeva 6, 10 000 Zagreb, Croatia

ABSTRACT

A four-year pot trial was set up to determine, as precisely as possible, the influence of increased levels of total petroleum hydrocarbons (TPH) upon soil and plants grown. In eight treatments, clean soil and different doses of drilling fluids and crude oil were applied. The changes in some chemical parameters of soil, plant density and crop yields were investigated. The influence of the studied indicators on the achieved plant density and crop yield was strongest in the first trial year. Drilling fluids had a stronger impact on the chemical properties of the studied soil, while plant density and yield were more strongly affected by crude oil. Upon application of drilling fluids and crude oil, the soil pH, contents of organic matter (OM) and heavy metals (HM) varied very little throughout the trial period, whereas the soil levels of total petroleum hydrocarbons, mineral oils (MO) and polycyclic aromatic hydrocarbons (PAHs) were significantly reduced after the first trial year.

INTRODUCTION

TPH introduction into the soil environment can occur from pipeline blow-outs, waste deposition after drilling oil and gas wells, road accidents, leaking underground storage tanks, land farming and uncontrolled landfill (Chaineau et al., 2003). With hindsight, it is interesting to note that it was once thought that a certain amount of TPH could serve as fertilizer and stimulate plant and animal growth (Carr, 1919, Murphy and Riley, 1929 and Mackin, 1950). The reasons for reduced plant growth in soils contaminated by TPH range from direct toxic effects of oil on plants (Baker, 1970 and Kyung-Hwa et al., 2004), lack of germination due to the lack of viable seeds (Ekundayo et al., 2001 and Ogboghodo et al., 2004a), reduced germination (Dorn and Salanitro, 2000), and unsatisfactory soil conditions. Soil conditions may be poor due to insufficient aeration caused by decreased air-filled pore space, and increased oxygen demand caused by oil-decomposing microorganisms, as well as a reduction in the level of available plant nutrients) (De Jong, 1980). According to available literature, the effects of TPH on soil used to be studied in two different ways. One investigation involved the addition of certain amounts of TPH into clean soil after sowing (Sarkar et al., 2005, Okolo et al., 2005 and Agbogidi et al., 2007). In the second investigation, the studied crop was sown into previously TPH contaminated soil (De Jong, 1980, Akaninwor et al., 2007 and Shahriari et al., 2007). This paper presents the results of experiment made by mixing the clean soil (treatment I) with a certain percentage of drilling fluids (treatments: II, III, IV and V) and then sowing the crops studied. The Experiment also included treatments in which clean soil was mixed with crude oil (treatments VII and VIII) taken from oil wells in the immediate vicinity of a pipeline breakage. Treatment VI represents soil that was hauled as "replacement" soil to the site of the pipeline breakage.

MATERIAL AND METHODS

A pot trial was set up in the greenhouse of the Faculty of Agriculture in Zagreb, Croatia, in the autumn 2003. The trial included 4 replications of each following treatments:

- Control (clean soil) taken in the immediate vicinity of a pipeline breakage site
- Drilling fluids — taken from the central waste pit of the oil/gas field
- 1/2 clean soil + 1/2 drilling fluids (6 kg clean soil + 6 kg drilling fluids)
- 2/3 clean soil + 1/3 drilling fluids (8 kg clean soil + 4 kg drilling fluids)
- 3/4 clean soil + 1/4 drilling fluids (9 kg clean soil + 3 kg drilling fluids)
- Soil hauled to the pipeline breakage site
- 2/3 clean soil + 1/3 crude oil (8 kg clean soil + 4 l crude oil taken from the oil well)

- 3/4 clean soil + 1/4 crude oil (9 kg clean soil + 3 l crude oil taken from the oil well)

The Experiment ran from 2003 to 2007. After the trial was prepared according to the above methodology, the crop sowing began. According to the crop sequence, winter wheat (Triticum aestivum L.) was sown on 14 October 2003 and 26 October 2005; winter barley (Hordeum vulgare L.) was sown on 21 October 2004 and 27 October 2006; soybean (Glycine hyspida L.) was sown on 29 June 2005 and 3 July 2006. The greenhouse trial was set up in pots with 4 replications; the trial pot area was 0.15 m². Standard agro-technical practices were applied to the pots — chemical protection and fertilization of crops. Basic soil chemical analyses (soil pH and organic matter content), TPH and MO contents, were performed once a year. HM and PAHs were determined in 2003, 2005, and 2007. Soil samples (for above mentioned analyses) were taken after crop harvest and prior to sowing the next crop.

The research objective was to investigate the possibility of field crops production on TPH contaminated soils, and to identify the effect of such contamination on plant density and crop yield by determining:

- changes in soil chemical complex (soil pH, OM, TPH and MO, HM and PAHs),
- Effects of different TPH and PAHs levels on emergence, plant density and yield of crops grown.

The investigation results were statistically processed by ANOVA and the t-test to estimate the significance of the differences between the treatments and control. The methods used to determine the studied parameters are given in Table 1.

Table 1: Methods used in investigations

Analysis	Method
Soil sampling	ISO 10381-1-8 (2001–2006)
Preparation of soil samples for physical and chemical analyses	ISO 11464:2004

Pretreatment of samples for determination of organic contaminants	ISO 14507:2003
Preparation of laboratory samples from large samples	ISO/DIS 23909:2007
Determination of organic (TOCIOM) and total carbon (TC) by dry digestion (elemental analysis)	ISO 10694:2004
Determination of total nitrogen by dry digestion (elemental analysis)	ISO 13878:2004
Determination of pH values (KCl) 1:2.5	ISO 10390:2004
Extraction of aqua regia soluble elements	ISO 11466:2004
Determination of Zn, Pb, Cd, Co, Ni, Cr and Cu using AAS	ISO 11047:2004 ISO 11885:1998
Determination of As, Ba, Mo, V and Hg using ICPMS	ISO/DIS 22036:2006
Determination of total petroleum hydrocarbons and mineral oils in soil --gas chmmatography	ISO 16703:2004
Determination of polycyclic aromatic hydrocarbons	ISO 18287:2005 EPA 550

RESULTS AND DISCUSSION

The major physical and chemical characteristics of water-based drilling fluids (muds) and crude oil applied in the trial are given in Table 2 and Table 3.

Table 2: Content of heavy metals in drilling fluids applied in the trial[a]

	Cd,	Hg,	Pb,	As,	Ni,	Cu,	Cr,	Zn,	Ha,	Ca,
	mg/kg	mg/kg	mg/kg	mg/kg	mg/kg	mg/kg	mg/kg	m g / kg	mg/kg	g/kg
Min	8.8	2.6	187	35.8	27.5	26.8	47.2	139	1988	6.85
Max	11.0	4.8	358	49.0	39.5	41.6	68.2	295	2674	11
Average	9.6	3.8	219	41.2	34.3	31.2	57.8	206	2373	9.03

[a]Survey of 20 drilling fluids samples taken from the central waste pit in the 011 field where pipeline breakage occurred.

Table 3: Some characteristics of crude oil applied in the trial[a]

	Sum, vol. %	Density at 15 °C, g/cm³	Viscosity at 37.8 °C mm²/s
Light gasoline	3.87	0.6702	
Light gasoline + heavy gasoline	19.27	0.7447	
Kerosene	9.53	0.8151	
Gas oil	14.28	0.8446	
High viscosity lubricant oil	9.94	0.8548–0.8705	7.5–20.6
Medium viscosity lubricant oil	6.88	0.8705–0.8832	20.6–43
High viscosity lubricant oil			Over 43
Residue	39.17	0.9530	
Loss	0.93		

Chromatographic analysis of fraction to 155 °C				
Paraffins	Naphthenes	Aromatics	Olefins	Sum
53.34	38.73	7.93		100.00

Metals, mg/kg								
Na–	Ca–	Mg–	Ni–	Pb–	Fe–	Zn–	Cr–	v–
0	12.94	0.44	6.31	0	1.04	0.49	0.24	0

[a]Average value of crude oil taken from the well near pipeline breakage.

Changes in Soil pH, Content of Organic Matter, Total Carbon, Total Nitrogen and C/N Ratio

Trial results show heterogeneity in soil pH and organic matter content. Table 4 and Fig. 1 present statistically significant differences in soil pH in all treatments, compared to control treatment (clean soil — treatment I). An increase in pH was recorded in treatments involving application of drilling fluids (100% or some other percent), which was expected because drilling fluids are rich in calcium (Table 2). Increased calcium levels in drilling fluids are a direct result of the use of calcium as an additive for preventing corrosion of oil/gas pumping pipes and for raising fluid density during drilling (Carls et al., 1995 and Bauder et al., 2005). $CaCO_3$ is also commonly added to bind fluids during remediation of pipeline breakages or other incidents. In treatments where crude oil was applied, pH changed statistically significantly compared to control treatment. At the pipeline breakage site (treatment VI), where clean soil was brought, pH also increased as a consequence of the different pH value of the hauled soil. Similar and statistically significant differences were recorded for OM (Table 4 and Fig. 2). The highest organic matter content was recorded in treatment II (100% drilling fluids). In other treatments, in accordance with the levels of drilling fluids added, statistically significantly higher OM values were recorded compared to the control treatment. These findings are attributed to the chemical composition of drilling fluids and crude oil. The analytical method of organic matter determination is based on the Dumas method of dry digestion, in which total carbon is determined in the sample, and then total organic carbon is determined upon treatment with dilute HCl solution. Multiplying TOC (Total Organic Carbon) by factor 1.85 gives OM; in other words, the OM percentage is obtained by the analysis of TOC, which naturally contains TPH. For this reason, the statistically significantly lower content of TOC in control treatment compared to all other treatments is not surprising (Table 4 and Fig. 3).

Table 4: Changes in soil pH, organic matter, total carbon, total nitrogen and C/N ratio

Treatment year	I	II	III	IV	V	VI	VII	VIII
Soil pH								

2003	6.51	6.67**	6.82**	6.78**	6.43**	7.13**	6.51	6.42**
2005	6.47	6.93**	6.89**	6.89**	6.64**	7.07**	6.54**	6.44**
2007	6.17	7.08**	7.03**	7.12**	6.75**	7.38**	6.71**	6.21**
Organic matter, %								
2003	2.13	5.57**	4.18**	3.24**	3.64**	3.54**	1.92**	2.23*
2005	2.43	5.39**	4.05**	3.44**	3.54**	3.44**	1.82**	2.33*
2007	2.33	5.31**	3.74**	3.54**	3.44**	3.24**	2.05**	3.44**
Carbon, %								
2003	1.74	4.90**	2.94**	2.62**	2.04**	1.96**	1.90**	1.83**
2005	1.67	4.85**	2.71**	2.56**	2.12**	1.81**	1.85**	1.91**
2007	1.69	4.89**	2.64**	2.57**	2.16**	1.88**	1.91**	1.80**
Nitrogen,%								
2003	0.22	0.18	0.17	0.16	0.17	0.12	0.14	0.15
2005	0.24	0.21	0.18	0.18	0.19	0.13	0.15	0.16
2007	0.21	0.19	0.19	0.18	0.20	0.15	0.15	0.17
Carbon/nitrogen ratio								
2003	8	27**	17**	16**	12**	16**	14**	12**
2005	7	23**	15**	14**	11**	14**	12**	12**
2007	8	26**	14**	14**	11**	13**	13**	11**

*, ** Significant at the 0.05 and 0.01 levels of probability respectively.

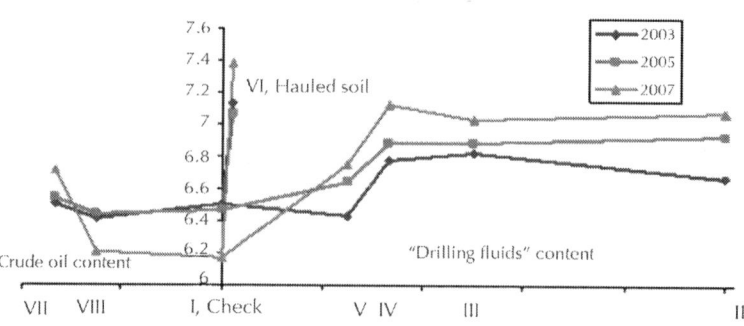

Figure 1: Effects of drilling fluids and crude oil on soil pH.

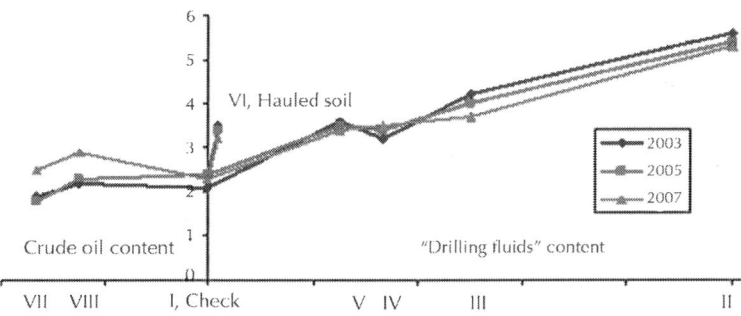

Figure 2: Effects of drilling fluids and crude oil on content of organic matter in soil (OM), %.

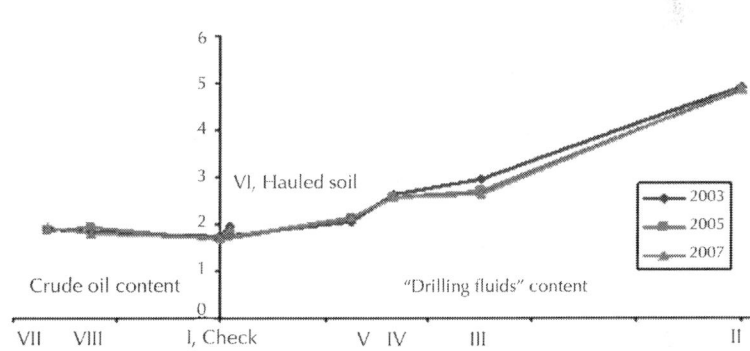

Figure 3: Effects of drilling fluids and crude oil on total carbon content in soil (TC), %.

Due to elevated carbon content and lower nitrogen content (Table 4) in treatments where drilling fluids were applied (treatments II and III), significant changes also occurred in the C/N ratio. The decrease in total nitrogen (Table 4 and Fig. 4) with an increase in TPH, MO (Table 5) and PAHs (Table 6) may be due to temporal immobilization of this nutrient by microbes, which might have increased in population. Jobson et al. (1974) and De Jong (1980) report on unfavourable carbon to nitrogen ratios in their investigations. Nitrogen addition with mineral fertilizers (Kirkpatrick et al., 2006) or organic soil improvers (Callaham et al.,

2002, Ogboghodo et al., 2004b and Adedokun and Ataga, 2007) may enhance the development of microbiological processes in soil and thereby improve the C/N ratio (Fig. 5).

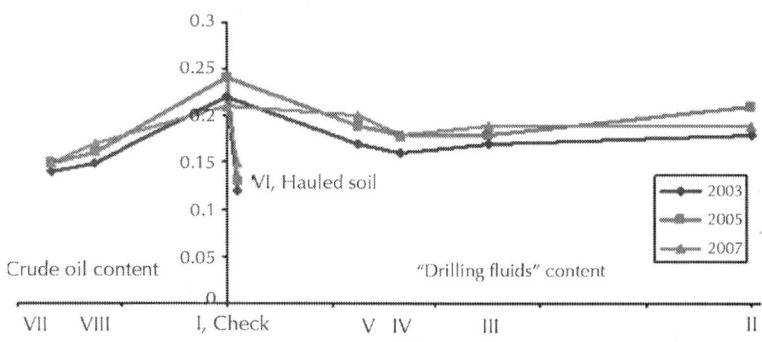

Figure 4: Effects of drilling fluids and crude oil on total nitrogen content in soil (TN), %.

Table 5: Changes in total petroleum hydrocarbon (TPH) and mineral oils (MO)

Treatment 1 year	I	II	III	IV	V	VI	VII	VIII
Total petroleum hydrocarbons-TPH, g kg^{-1}								
2003	0.44	76.1**	24.5**	14.7**	6.3**	1.1**	2.4**	3.1**
2005	0.06	8.0*	4.1**	2.9**	2.2**	3.8**	1.8**	2.5**
2007	0.08	5.9**	3.5**	2.1**	1.6**	2.7**	1.3**	2.2**
Mineral oils — MO, g kg^{-1}								
2003	0.23	49.6**	7.4**	4.2**	3.5**	0.4**	0.6**	0.4**
2005	0.01	1.6**	1.1**	0.9**	0.2**	1.5**	0.2**	0.8**
2007	0.03	1.7**	1.1**	0.7**	1.6**	1.2**	0.8**	0.**

*, ** Significant at the 0.05 and 0.01 levels of probability respectively.

Table 6: Soil contamination by polycyclic aromatic hydrocarbons (PAHs)

Type of PAHs	Four-ring, mg kg⁻¹ dry soil				Five- or six-ring, mg kg⁻¹ dry soil				Total
	Flour-anthene	Pyrene	Benzo (b) fluoranthene	Benzo (k) fluoranthene	Benzo (a) pyrene	Dibenzo (ah) anthracene	Benzo (ghi) perylene	Indeno (1,2,3-cd) pyrene	
Treatment									
2003									
I	< 0.01	< 0.01	< 0.01	< 0.01	< 0.01	< 0.01	< 0.01	< 0.01	< 0.01
II	5	< 0.01	42	< 0.01	< 0.01	1.5	< 0.01	< 0.01	48.50**
III	6	8	< 0.01	< 0.01	< 0.01	< 0.01	< 0.01	< 0.01	14.00**
IV	< 0.01	< 0.01	< 0.01	< 0.01	2.5	3.2	4.1	0.5	10.30**
V	1	1.1	< 0.01	< 0.01	2.1	1.5	1.4	1.4	8.50**
VI	0.1	0.1	0.1	0.1	0.9	0.1	014	0.1	1.64**
VII	8.1	5.9	2.4	4.7	9.1	4.7	5.7	3.1	43.70**
VIII	6.7	4.1	2.7	3.8	10.1	5.4	6.1	2.8	41.70**
2005									
I	< 0.01	< 0.01	< 0.01	< 0.01	< 0.01	< 0.01	< 0.01	< 0.01	< 0.01
II	0.3	< 0.01	4.1	< 0.01	< 0.01	0.2	< 0.01	< 0.01	4.6**
III	< 0.01	< 0.01	0.03	< 0.01	< 0.01	< 0.01	< 0.01	< 0.01	0.03**
IV	< 0.01	< 0.01	< 0.01	< 0.01	< 0.01	< 0.01	< 0.01	< 0.01	< 0.01
V	< 0.01	< 0.01	0.03	0.02	< 0.01	< 0.01	< 0.01	< 0.01	0.05**

VI	< 0.01	< 0.01	0.01	0.01	< 0.01	< 0.01	< 0.01	< 0.01	0.02*
VII	0.4	< 0.01	0.2	0.5	0.8	< 0.01	0.1	0.1	2.1**
VIII	0.4	< 0.01	0.9	0.1	1.7	0.1	0.1	< 0.01	3.3**
2007									
I	< 0.01	< 0.01	< 0.01	< 0.01	< 0.01	< 0.01	< 0.01	< 0.01	< 0.01
II	< 0.01	< 0.01	< 0.01	< 0.01	< 0.01	< 0.01	< 0.01	< 0.01	< 0.01
III	< 0.01	< 0.01	< 0.01	< 0.01	< 0.01	< 0.01	< 0.01	< 0.01	< 0.01
IV	0.01	< 0.01	0.03	0.02	< 0.01	< 0.01	< 0.01	< 0.01	0.05**
V	< 0.01	< 0.01	< 0.01	< 0.01	< 0.01	< 0.01	< 0.01	< 0.01	< 0.01
VI	< 0.01	< 0.01	< 0.01	< 0.01	< 0.01	< 0.01	< 0.01	< 0.01	< 0.01
VII	< 0.01	< 0.01	< 0.01	< 0.01	< 0.01	< 0.01	< 0.01	< 0.01	0.02*
VIII	< 0.01	< 0.01	0.02	0.02	< 0.01	< 0.01	< 0.01	< 0.01	0.04**

*, ** Significant at the 0.05 and 0.01 levels of probability respectively.

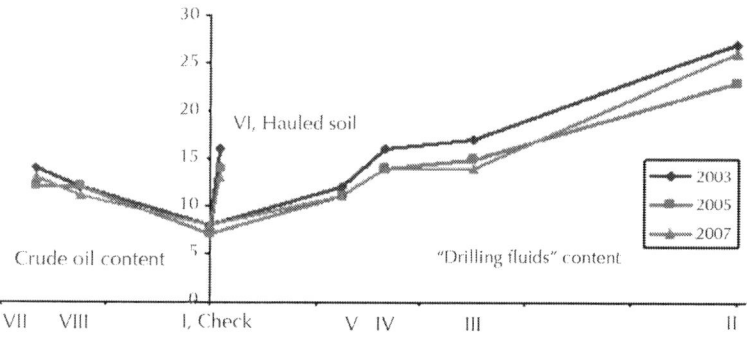

Figure 5: Effects of drilling fluids and crude oil on soil C/N ratio.

Total Petroleum Hydrocarbons and Mineral Oils

Crude oil and petroleum products are complex mixtures of hundreds of hydrocarbon compounds, ranging from light, volatile, short-chained organic compounds to heavy, long-chained, branched compounds. Soil contamination by TPH and MO is a growing concern since it may be a source of groundwater contamination (Healy et al., 2001 and Asia et al., 2007). Contaminated soils can reduce the usability of land (Carls et al., 1995), and weathered petroleum residuals may remain bound to soils for years (Sarkar et al., 2005). MO contains hundreds of hydrocarbon compounds, including a substantial fraction of nitrogen and sulphur-containing compounds. Hydrocarbons are mainly mixtures of straight and branched chain hydrocarbons (alkanes), cycloalkanes and aromatic hydrocarbons.

Soil contamination by TPH and MO is illustrated in Table 5 and Fig. 6 and Fig. 7. Compared to the control, statistically higher levels of TPH and MO were found in all treatments during the trial period. This was most significant in the first trial year. In the second, third, and fourth years, the TPH and MO contents decreased in relation to the first year, but they were still (statistically) significantly higher compared to the control. Residual TPH and MO degraded very slowly after the first year and their levels remained constant in the remaining trial period. These findings indicate that in the first trial year, most of the TPH and MO, with the volatile aromatic fraction prevalent in their composition (Table 3), evaporate into the air or degrade through microbiological processes (Jobson et al., 1974, Yong et al., 1992, Chaineau et al., 2003 and Ogboghodo et al., 2004b). Depending on the soil's mechanical composition, TPH and MO bind strongly to the soil adsorption complex, and only slightly and slowly percolate towards deeper horizons (Rhykerd et al., 1999 and Pezeshki et al., 2000). TPH and MO left over in the soil in amounts below 5 g/kg of soil, degrade very slowly and, according to the available research results, are not harmful to crops. More precisely, the studied crops grow without much greater restraint on soil containing less TPH and MO than 5 g/kg soil.Akaninwor et al. (2007) and Chaineau et al. (2003) came to similar conclusions in their investigations.Kyung-Hwa et al. (2004) report that 10 g/kg soil was toxic to the studied crops, but also that no phytotoxicity was determined in soil containing up to 1 g/kg hydrocarbons. Our previous investigations (Kisic et al., 2005) have shown that the rate of TPH and MO degradation in soil was mainly connected with the kind, quantity, and type of crude oil or drilling fluid, OM content and soil mechanical composition (clayey or sandy soil), soil moisture, together with the season in which increased TPH content was spilled into the soil.

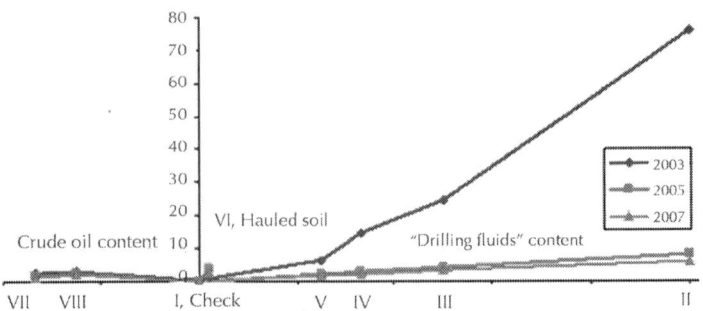

Figure 6: Effects of drilling fluids and crude oil on total petroleum hydrocarbons (TPH) in soil, g/kg.

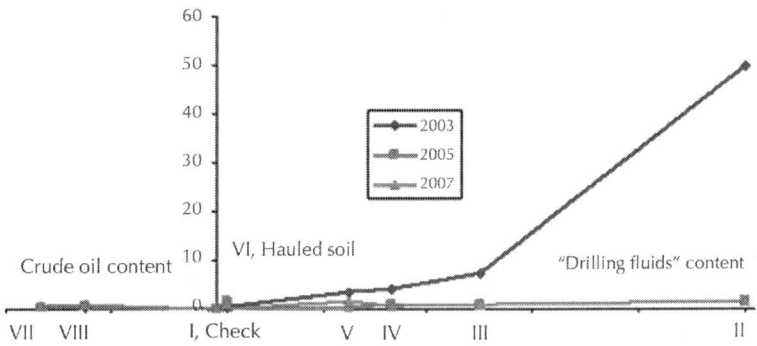

Figure 7: Effects of drilling fluids and crude oil on mineral oil (MO) in soil, g/kg.

Polycyclic Aromatic Hydrocarbons

Crude oil and drilling fluids are a complex mixture of hydrocarbons containing PAHs and non-hydrocarbon(s) compounds including HM, which are potentially phytotoxic (Kelsey and Alexander, 1997 and Samanta et al., 2002), and may interfere with normal plant development and reproduction (Mendelssohn et al., 1990 and Adam and Duncan, 2002). PAHs are large group of polycyclic hydrocarbons containing one or more benzene rings. According to the World Health Organization's (WHO) recommendations for consideration and interpretation, measurements of emission into the atmosphere should involve only benzo(a)pyrene, as the best known and most studied among several thousands of PAHs, whose toxic, carcinogenic and mutagenic action is evidence-based (WHO, 2000). Environmental contamination by PAHs, especially PAHs of high molecular mass, primarily refers to contamination of the atmosphere. PAHs with a smaller number of rings exist in the atmosphere in gaseous form, while those having a larger number of rings are adsorbed by air particles (Lee et al., 1981). Water and soil contamination by PAHs is regarded as secondary contamination since the air-borne PAHs are bound to suspended particles, and are deposited onto soil and water (Tritscher, 2004).

Table 6 and Fig. 8 show the PAHs status in soil in particular trial years. In the first trial year, the highest PAHs content was recorded in treatment II, and in treatments VII and VIII where crude oil was applied. Although the PAHs sum was almost equal in treatments II, VII and VIII, differences in particular PAHs are discernible between the treatments. In treatment II, due to its chemical composition and origin, benzo(b)fluorantene absolutely prevails in the PAHs sum, while all the 8 studied PAHs are equally represented in treatments VII and VIII where crude oil was applied (Table 6). This is attributed to the high volatility of PAHs (Samsøe-Petersen et al., 2002; Serrano et al., 2006). In treatments VII and VIII, where crude oil was applied, PAHs were not yet degraded by physical, chemical or biological processes. Owing to its total higher levels in soil compared to other PAHs, benzo(b)fluorantene remains longer in soil. At the time of repeated sampling in 2005, there were almost no PAHs in the soil; increased benzo (b)fluorantene was found only in treatment II (Table 6) compared to all the other treatments. At the last sampling in 2007, PAH presence in soil was insignificant because they were partially degraded by physical processes (Fismes et al., 2002 and Samsøe-Petersen et al., 2002), by microorganisms (Samanta et al., 2002), or some other chemical or biological process (Riser-Roberts, 1998).

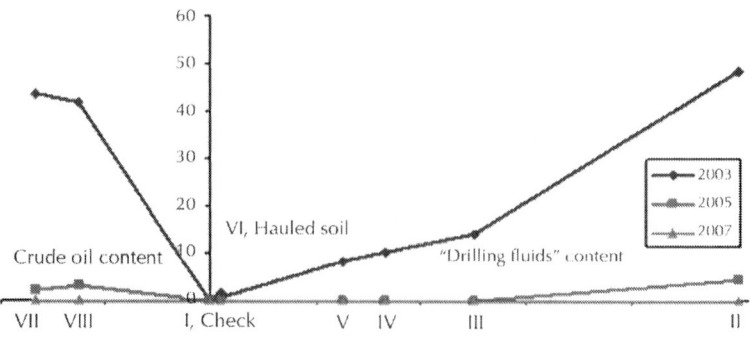

Figure 8: Effects of drilling fluids and crude oil on total Polycyclic aromatic hydrocarbons (PAHs) in soil, mg/kg.

Heavy Metals

Of the numerous impacts on soil quality, a significant part is contributed by HM (Colins et al., 2006). This concept implies the presence of HM in soil in amounts causing a visible or measurable disturbance of some of the soil functions (Abrahams, 2002 and McLaren, 2003).

As soil is a very complex system encompassing numerous processes, some of which are irreversible and plants are subject to diverse biotic and abiotic influences, it is not always simple to assess soil and plant interactions and synergistic action, due to the occurrence of changes that would allow classifying the soil as contaminated. For these reasons, soil contamination is significantly different from (for example) air or water contamination. It is likewise very difficult to speak about threshold values of soil contamination by HM. Some values are applicable to heavy clay soils, whilst other values will hold for lighter sandy soils; some values will apply to soils used in agriculture, while other values will hold for soils located in various industrial or urban zones (Abrahams, 2002 and Samsøe-Petersen et al., 2002).

Table 7 shows the status of soil contamination by certain HM (Cd, Hg, Pb, Mo, As, Ni, Co, Cu, Cr, Zn, Ba and V). The lowest HM content was in all cases determined in the control treatment. In all other treatments, changes in HM levels were conditioned by the degree of soil contamination. It can be seen from Table 7that cadmium, lead, molybdenum, nickel, cobalt and chromium contents did not differ significantly between the trial years. Mercury, arsenic, copper, zinc, barium and vanadium contents differed statistically between the trial years according to the degree of soil contamination. The highest mercury, arsenic, copper, zinc, barium and vanadium contents were determined in treatment II. This is associated with the chemical composition of drilling fluids applied in the trial (Table 2). As increased levels of mercury, arsenic, zinc and barium were found in these fluids, raised levels of the listed metals could be expected in treatments involving application of a certain percentage of drilling fluids. It is important to note that crude oil contains neither zinc nor barium; their increased presence in drilling fluids is attributed to the use of zinc carbonate and barite (barium sulphate) in oil drilling (Carls et al., 1995 and Scholten et al., 2000). Barium is insoluble, inert and non-toxic (Monaghan et al., 1980), so it is not considered a serious problem in soil. However, mercury, arsenic and zinc require quite a different approach. Higher levels of these metals in soil call for recultivation through electroremediation (Riser-Roberts, 1998), phytoremediation (Hazel, 2005) or solidification (Asia et al., 2007). Approximately equal copper and zinc contents were found in the treatment with drilling fluids, as well as in the treatment in which the contaminated soil was replaced by soil hauled to the site. This indicates the necessity to assess the quality of the soil to be transported to a pipeline breakage site.

Table 7: Total heavy metal contents in soil, mg/kg

	Treatment year	I	II	III	IV	V	VI	VII	VIII
Cadmium	2003	0.33	0.23	0.24	0.29	021	0.32	021	0.37
	2005	0.32	0.46	043	0.32	0.33	0.43	0.27	0.35
	2007	0.37	0.38	0.38	0.36	0.36	0.32	0.35	0.33
Mercury	2003	0.02	0.18**	0.03*	0.08**	0.04*	0.06**	0.02	0.09**
	2005	0.02	0.26**	0.15**	0.08**	0.04**	0.05**	0.04**	0.03*
	2007	0.01	0.16**	0.09**	0.07**	0.05**	0.07**	0.03*	0.04**
Lead	2003	13	18	14	19	15	14	14	12
	2005	13	18	18	15	16	15	20	15
	2007	15	26	21	20	20	31	17	19
Molybde-num	2003	0.5	0.4	0.4	0.5	0.7	0.6	0.4	0.4
	2005	0.6	0.4	0.6	0.6	0.4	0.8	0.4	0.6
	2007	0.7	1.0	1.1	1.1	1.0	1.1	1.0	1.0
	2003	7	18**	14**	16**	9*	9*	9*	10**
	2005	6	32**	16**	19**	11**	11**	10**	8*
	2007	7	33**	15**	16**	9*	31**	7	8*
Nickel	2003	17	25	23	28	20	20	20	19
	2005	17	15	18	15	15	15	17	14
	2007	12	14	14	13	13	14	12	12

Cobalt	2003	7	9	10	12	10	9	8	6
	2005	7	7	6	9	9	8	9	9
	2007	12	11	12	14	11	11	12	13
Copper	2003	15	18**	16	15	12**	18**	15	14
	2005	14	20**	18**	14	12**	19**	14	10**
	2007	14	22**	18**	17*	15	21**	14	14
Chro-mium	2003	40	30	25	55	25	50	22	22
	2005	40	45	26	27	24	30	25	16
	2007	30	45	38	36	33	44	31	31
Zinc	2003	70	78**	65	69	70	67*	55**	63**
	2005	70	81**	61	65*	70	71	50**	45**
	2007	51	92**	74**	70**	62*	91**	54*	58**
Barium	2003	79	1872**	1356**	1423**	709**	195**	419**	97**
	2005	75	1936**	1228**	1451**	634**	202**	412**	101**
	2007	80	2000**	1500**	1500**	800**	500**	400**	101**
Vana-dium	2003	25	32**	32**	40**	25	22**	29**	19**
	2005	25	21**	22**	22**	25	21**	29**	24
	2007	30	25**	28*	24**	28*	24**	32*	30

*, ** Significant at the 0.05 and 0.01 levels of probability respectively.

Changes in Plant Density and Yield of Crops Grown

In the first trial year, as early as in the germination/emergence stage, differences were observed in the emergence of winter wheat (cultivar: Zlatni dukat) depending on the different TPH levels in the soil. Winter wheat emergence was inversely proportional to TPH levels in the soil. In the treatments with higher soil contamination (treatments II; IV; VII and VIII), crop emergence was much poorer compared to treatments III and V with lower TPH content. Higher contamination caused formation of a thin film around the seed germ and thereby prevented the inflow of oxygen, which caused embryo death. Another reason for poorer emergence is the fact that TPH contaminated soil becomes more compact and less moist, and has a higher content of toxic substances. Ferrell et al. (1984), Issoufi et al. (2006), Shahriari et al. (2007), Adedokun and Ataga (2007) also report that oil pollution inhibits seed germination and plant growth. If the number of plants determined in the control treatment (100%) is taken as optimal, it can be seen that plant density at the wheat harvest in treatment II was 74% (that) of the control, in treatment III — 98%, in treatment IV — 60%, in treatment V — 88%, while the plant density in treatment VI was only 50% (that) of the control (Table 8). Plant density in treatments VII and VIII was 57% and 37% (that) of the control, respectively. This indicates that crude oil had a much stronger effect on the achieved plant density than the drilling fluids. Plant density achieved in the control treatment was statistically higher compared to other treatments, whereas no statistically significant differences were determined between other treatments (Table 8).

Table 8: Yield, plant density and some components of winter wheat

Treatment/ year/compo- nents	I	II	III	IV	V	VI	VII	VIII
2003/04								
Yield, g pot⁻¹	44.5	22.3**	31.7**	26.9**	28.0**	26.9**	22.4**	23.6**
Total plants	106	79	104	63**	93	53**	60**	39**
Plants with ears	71	63	74	47**	60	43**	48**	37**
Plants with- out ears	35	16**	30	16**	33	10**	12**	2**
2005/06								
Yield, g pot⁻¹	46.6	42.6	49.0	45.6	41.6	48.1	36.4**	37.8**
Total plants	129	86*	114	123	113	112	115	115
Plants with ears	108	81*	96	103	95	94	96	97
Plants with- out ears	21	5**	18**	20*	18**	18**	19**	18

*, ** Significant at the 0.05 and 0.01 levels of probability respectively.

According to the crop sequence, winter barley, cv. Rex was sown into pots in the second trial year. Even at the initial emergence stages, differences from the first trial year (when wheat was sown) were observable. The number of plants and the plant density achieved did not reflect the differences that were recorded during the first year. No statistically significant differences between trial treatments were determined with respect to the number of plants. The achieved yield, however, shows statistical differences, {but not as marked as in the first year} (Table 9). The highest difference was found between treatments I and II, and treatment VIII, while other treatments had more or less uniform yields.

Table 9: Yield, plant density and some components of winter barley

Treatment/ year/compo-nents	I	II	III	IV	V	VI	VII	VIII
2004/05								
Yield, g pot^{-1}	61.3	63.9	56.2*	54.2*	57.1	57.3	54.9	49.6**
Total plants	119	128	118	106	109	122	117	131
Plants with ears	114	128*	116	106	108	119	111	100*
Plants without ears	4.5	0**	2**	0**	1**	3**	6**	31**
2006/07								
Yield, g pot^{-1}	63.8	61.8	56.2*	54.2*	57.1*	57.4	55.0*	54.7*
Total plants	129	128	118	110	112	122	117	116
Plants with ears	124	124	116	106	109*	119	111	109*
Plants without ears	5	4**	2**	3**	3**	3**	6**	6**

*, ** Significant at the 0.05 and 0.01 levels of probability respectively.

After the barley was harvested, soybean, cv. Sabina (00-000 maturity group) was sown into pots. The achieved plant density and yield manifested very interesting changes compared to the preceding year when barley was grown. As usual, uniform seed rate was applied to all pots. However, essential differences in plant density and yield were already observed at the emergence stage, as well as in the growing period. The largest number of plants was recorded in the control treatment over the entire growing period, while the number of plants differed statistically in other treatments (Table 10). By far the lowest plant density was achieved in treatment VI, while that of other treatments was significantly lower than the control. The highest yield was obtained in the control treatment, while significantly lower yields were recorded in other treatments. Although three years had passed since the beginning of the trial, it was evident that TPH contamination exerted a significant effect on the number of emerged plants, and thereby also on the achieved soybean density.

Table 10: Yield, plant density and some components of Soybean

Treatment/ year/compo- nents	I	II	III	IV	V	VI	VII	VIII
2005								
Yield, g pot⁻¹	32.7	18.0**	19.2**	17.9**	20.3*	4.9**	23.2*	22.4*
Total plants	14	8**	6**	5**	9**	2**	8**	8**
Total pods	104	54**	51**	44**	73	13**	76	72
2006								
Yield, g pot⁻¹	14.6	11.1	14.3	13.4	11.6	15.9	16.7	16.3
Total plants	8	8	9	8	8	8	8	9
Total pods	37	35	39	34	40	46**	47**	53**

*, ** Significant at the 0.05 and 0.01 levels of probability respectively.

Winter wheat (the same cultivar as in the first year) was sown again in the autumn of 2005. Early in the initial emergence stages, changes in plant density were observed in comparison with the first trial year. Like in the first year, about 200 germinated seeds were sown per pot. Plant density in treatment II was 70% and in treatment IV 95% that of the control treatment. Plant density of other treatments was about 90% (that) of the control.

Soybean, this time cultivar Dora (0-00 maturity group), was sown again in July 2006. Compared to 2005, when cultivar Sabina was sown in pots, plant density was very different in 2006. This calls for an explanation of the difference in the number of plants that emerged in the control treatment in the two years when soybean was grown in pots (Table 10). Sabina is a soybean cultivar that is sown at a higher density (1,400,000 plants per hectare), whereas the sowing density of cultivar Dora is 650,000 plants per hectare. This accounts for the different number of plants that emerged in particular treatments. No statistically significant differences in the number of emerged plants were recorded in 2006, while the total number of pods in treatments VI, VII and VIII was statistically higher compared to the control. No statistically significant differences in yield were found between the studied treatments.

Upon soybean harvest in the autumn of 2006, winter barley was sown into pots, the same cultivar as in 2004. Data given in Table 9 show an approximately equal plant density at earing and at harvest in all treatments. There was no statistically significant difference between the control and treatments II and VI either. Significant differences, however, were recorded relative to the other treatments. Similar conclusions regarding plant density and yields achieved were reached by Shahriari et al. (2007), Akaninwor et al. (2007). They determined differences in yield and plant density between stubble crops and soybean as a spring row crop. Chaineau et al. (2003) report that the resistance of seeds to oil contamination followed the decreasing order: wheat > barley > maize > pea > lettuce.

The presented results of the four-year investigations and growing six crops indicate that the application of drilling fluids and crude oil leads to changes both in soil and in crops grown. These changes were most evident in the first trial year, with a marked decline in some parameters in the subsequent years. Upon application of the studied materials, soil pH, OM and HM contents remained constant throughout the trial period, whereas significant changes occurred in the case of TPH, MO and PAHs after the first trial year. As drilling fluids are rich in calcium and carbon, (only if contaminants (HM) were removed from drilling fluids) these materials could be used for liming acid soils as soil improvers only if contaminants (HM) were removed from them (drilling fluids) (Miller and Pesaran, 1980).

CONCLUSIONS

(I would write this in present SK) drilling fluids have (had) a stronger influence on the studied soil chemical properties while crude oil affected plant density and crop yield more strongly. The level of soil contamination by TPH and PAHs in the first trial year had a crucial role for the achieved plant density and yield. The raised content of TPH and PAHs in soil has the strongest influence in the emergence stage. TPH and PAHs in soil forms a thin film around the seed germ and thereby hampers oxygen inflow, leading to embryo death and/ or slower emergence of the plant.

TPH levels below 5 g/kg soil or 5 mg/kg PAHs in soil had no significant effect on the plant density of crops grown. Consequently, the value of 5 g/kg TPH and 5 mg/kg PAHs in soil could be recommended as a warning or emergency value in remediation of hydrocarbons-contaminated soil. The presented results show that the largest part of TPH and PAHs were lost through bioremediation (bioaugmentation and/or biostimulation) processes and soil mixing during the first and second trial years.On the other hand, the levels of heavy metals did not change much over the four years trial. This indicates that the problem of soil contamination by TPH, and partially the PAHs problem, may be solved through soil bioremediation or aeration by tilling practices (ploughing and harrowing). Increased content of heavy metals in soil requires quite a different approach to soil cleansing by electroremediation, phytoremediation or solidification. The extent of environmental consequences and changes in the soil chemical complex, plant density and yield of crops grown on TPH and PAHs contaminated soil depend primarily on the type and quantity of crude oil or drilling fluids unintentionally introduced into the environment, soil type, crops grown, the season, climate, and various other influences.

ACKNOWLEDGMENTS

This paper presents results of research programs supported by the Ministry of Science, Education and Sports of Republic Croatia.

REFERENCES

1. Abrahams, P.W., 2002. Soils: their implications to human health. The Science of TheTotal Environment 291 (1–3), 1–32.

2. Adam, G., Duncan, H., 2002. Influence of diesel fuel on seed germination. EnvironmentalPollution 120 (2), 363–370.

3. Adedokun, O.M., Ataga, A.E., 2007. Effects of amendments and bioaugumentation of soilpolluted with crude oil, automotive gasoline oil, and spent engine oil on the growthof cowpea (Vigna ungiculata L. Walp). Scientific Research and Essay 2 (5), 147–149.

4. Agbogidi, O.M., Eruotor, P.G., Akparobi, S.O., Nnaji, G.U., 2007. Evaluation of crude oilcontaminated soil on the mineral nutrient elements of maize (Zea Mays L.). Journalof Agronomy 6 (1), 188–193.

5. Akaninwor, J.O., Ayeleso, A.O., Monago, C.C., 2007. Effect of different concentrations ofcrude oil (Bonny light) on major food reserves in guinea corn during germinationand growth. Scientific Research and Essay 2 (4), 127–131.

6. Asia, I.O., Jegede, S.I., Jegede, D.A., Ize-Iyamu, O.K., Akpasubi, E.B., 2007. The effects ofpetroleum exploration and production operations on the heavy metals contents ofsoil and groundwater in the Niger Delta. Journal of Physical Science 2 (10), 271–275.

7. Baker, J.M., 1970. The effects of oils on plants. Environmental Pollution 1 (1), 27–44.

8. Bauder, T.A., Barbarick, K.A., Ippolito, J.A., Shanahan, J.F., Ayers, P.D., 2005. Soil propertiesaffecting wheat yields following drilling-fluid application. Journal of

EnvironmentalQuality 34, 1687–1696.

9.   Callaham, M.A., Stewart, A.J., Alarcon, C., McMillen, S.J., 2002. Effects of earthworm(Eisenia fetida) and wheat (Triticum aestivum) straw additions on selected propertiesof petroleum contaminated soils. Environmental Toxicology and Chemistry 21/8,1658–1663.

10. Carls, E.G., Dennis, B.F., Chaffey, S.A., 1995. Soil contamination by oil and gas drilling andproduction operations in Padre Island National Seashore, Texas, USA. Journal ofEnvironmental Managment 45, 273–286.

11. Carr, R.H., 1919. Vegetative growth in soils containing crude petroleum. Soil Science 8,67–68.

12. Chaineau, C.H., Yepremian, C., Vidalie, J.F., Ducreux, J., Ballerini, D., 2003. Bioremediationof a crude oil-polluted soil. Biodegradation, Leaching and Toxicity Assesments 144,419–440.

13. Colins, C., Fryer, M., Grosso, A., 2006. Plant uptake of non-ionic organic chemicals.Environmental Science and Technology 40, 45–52.De Jong, E., 1980. The effect of a crude oil spill on cereals. Environmental Pollution 22,187–196.

14. Dorn, P.B., Salanitro, J.P., 2000. Temporal ecological assesment of oil contaminated soilsbefore and after bioremediation. Chemosphere 40, 419–426.

15. Ekundayo, E.O., Emede, T.O., Osayande, D.I., 2001. Effects of crude oil spillage on growthand yield of maize (Zea mays L.) in soils of midwestern Nigeria. Plant Foods forHuman Nutrition 56 (4), 313–324.

16. Ferrell, R.E., Seneca, E.D., Linthurst, R.A., 1984. The effects of crude oil on the growth ofSpartina alterniflora Loisel and Spartina cynosuroides (L.) Roth. Journal of ExperimentalMarine Biology and Ecology 83 (1), 27–39.

17. Fismes, J., Perrin-Ganier, C., Empereur-Bissonet, P., Morel, J.L., 2002. Soil-to-root transferand translocation of polycyclic aromatic hydrocarbons by vegetables grown onindustrial contaminated soils. Journal of Environmental Quality 31,

1649–1656.

18. Hazel,W., 2005. Suck it up. Phytoremediation. Available at http://ourgardengang.tripod.com.

19. Healy, M., Wise, D.L., Moo-Young, M., 2001. Environmental Monitoring and Biodiagnosticsof Hazardous Contaminants. Kluwer Academic Publishers, Boston-London.

20. Issoufi, I., Rhykerd, R.L., Smiciklas, K.D., 2006. Seedling growth of agronomic crops incrude oil contaminated soil. Journal of Agronomy and Crop Science 192, 310–317.

21. Jobson, A., McLaughlin, M., Cook, F.D., Wstlake, D.W.S., 1974. Effects of amendments onthe microbial utilisation of oil applied to soil. Journal of Applied Microbiology 27,166–171.

22. Kelsey, J.W., Alexander, M., 1997. Declining bioavalibility and inappropriate estimationof risk of persistent compounds. Environmental Toxicology and Chemistry 16,582–585.

23. Kirkpatrick, W.D., White Jr., P.M., Wolf, D.C., Thoma, G.J., Reynolds, C.M., 2006. Selectingplants and nitrogen rates to vegetate crude-oil-contaminated soil. InternationalJournal of Phytoremediation 8 (4), 285–297.

24. Kisic, I., Basic, F., Mesic, M., Veronek, B., Vadjic, Z., Mesic, S., 2005. Changes in soil andcrop yield caused by oil incidents. Cereal Research Communications 33 (1),243–246.

25. Kyung-Hwa, B., Hee-Sik, K., Hee-Mock, O., Byung-Dae, Y., Jaisoo, K., In-Sook, L., 2004.Effect of crude oil, oil components, and bioremediation on plant growth. Journal ofEnvironmental Science and Health A39 (9), 2465–2472.

26. Lee, M.L., Novotny, M.V., Bartle, K.D., 1981. Analytical chemistry of Polycyclic AromaticCompounds. Academic Press, New York.

27. Mackin, J.G., 1950. Report on a study of the effects of application of crude petroleum onsaltgrass (Distichlis spicata L.) Green. Texas A&M Research Foundation.

28. McLaren, R.G., 2003. Micronutrients and toxic elements. In: Benbi, D.K., Nieder, R. (Eds.),Handbook of processes and

modeling in the Soil-Plant System. The Haworth Press,New York, pp. 589–618.

29.   Mendelssohn, I.A., Hester, M.W., Sasser, C., Fishel, M., 1990. The effect of a Lousianacrude oil discharge from a pipeline break on the vegetation of a Southeast Lousianabrackish marsh. Oil and chemical pollution 7 (1), 1–15.

30.   Miller, R.W., Pesaran, P., 1980. Effects of drilling fluids on soil and plants. II Completedrilling fluid mixtures. Journal of Environmental Quality 9, 552–556.

31.   Monaghan, P.H., McAuliffe, C.D., Weis, 1980. Environmental aspects of drilling fluids andcuttings from oil and gas operation in offshore and coastal waters. MarineEnvironmental Pollution 1: Hydrocarbons. Elsevier Scientific.

32.   Murphy, J.F., Riley, J.P., 1929. Some effects on crude petroleum on nitrate production,seed germination and growth. Soil Science 24, 117–120.

33.   Ogboghodo, I.A., Iruaga, E.K., Osemwota, I.O., Chokor, J.U., 2004a. An assessment of theeffects of crude oil pollution on soil properties, germination and growth of maize(Zea Mays) using two crude types — Forcados Light and Escravos Light. Environmental Monitoring and Assessment 96, 143–152.

34.   Ogboghodo, I.A., Erebor, E.B., Osemwota, I.O., Isitekhale, H.H., 2004b. The effects ofapplication of poultry manure to crude oil polluted soils on maize growth and soilproperties. Environmental Monitoring and Assessement 96 (1–3), 153–161.

35.   Okolo, J.C., Amadi, E.N., Odu, C.T.I., 2005. Effects of soil treatments containing poultrymanure on crude oil degradation in a sandy loam soil. Applied Ecology andEnvironmental Research 3 (1), 47–53.

36.   Pezeshki, S.R., Hester, M.W., Lin, Q., Nyman, J.A., 2000. The effects of oil spill and cleanupon dominant US Gulf coast marsh macrophytes: a review. EnvironmentalPollution 108 (2), 129–139.

37. Rhykerd, R.L., Crews, B., McInnes, K.J., Weaver, R.W., 1999. Impact of bulking agents,forced aeration, and tillage on remediation of oil-contaminated soil. BioresourceTechnology 67 (3), 279–285.

38. Riser-Roberts, E., 1998. Biodegradation/mineralization/ biotransformation/bioaccumulationof petroleum. Remediation of petroleum contaminated soils — Biological,Physical and Chemical Processes. Lewis Publishers, Florida, pp. 115–192.

39. Samanta, S.K., Singh, O.V., Jain, R.K., 2002. Polycyclic aromatic hydrocarbons:environmental pollution and bioremediation. Trends in Biotechnology 20 (6),243–248.

40. Samsøe-Petersen, L., Larsen, H.E., Larsen, B.P., Brun, P., 2002. Uptake of trace elementsand PAHs by fruit and vegetables from contaminated soils. Environmental Scienceand Technology 36, 3057–3063.

41. Sarkar, D., Ferguson, M., Datta, R., Birnbaum, S., 2005. Bioremediation of petroleumhydrocarbons in contaminated soils: comparison of biosolids addition, carbonsupplementation, and monitored natural attenuation. Environmental Pollution 136 (1),187–195.

42. Scholten, M.C.Th., Karman, C.C., Huwer, S., 2000. Exotoxicological risk assesment relatedto chemicals and pollutants in off-shore oil production. Toxicology Letters 112–113,283–288.

43. Serrano, A., Gallego, M., González, J.L., 2006. Assessment of natural attenuation ofvolatile aromatic hydrocarbons in agricultural soil contaminated with diesel fuel.Environmental Pollution 144 (1), 203–209.

44. Shahriari, M.H., Savaghebi-Firoozabadi, G., Azizi, M., Kalantari, F., Minai-Tehrani, D.,2007. Study of growth and germination of Medicago Sativa (Alfalfa) in light crudeoil-contaminated soil. Research Journal of Agriculture and Biological Sciences 3 (1),46–51.

45. Tritscher, A.M., 2004. Human health risk assessment of

processing-related compoundsin food. Toxicology Letters 149, 177–186.

46.   WHO, 2000. Polycyclic Arpomatic Hydrocarbons (PAHs). Air Quality Guidelines. WHO,Copenhagen, Denmark.

47.   Yong, R., Mohamed, A.M.O., Warkentin, 1992. Principles of Contaminant Transport inSoils. Elsevier Science Publishers, Amsterdam, p. 328.

Influence of Viscosity Modifier Nature and Concentration on the Viscous Flow Behaviour of Oil-based Drilling Fluids at High Pressure

J. Hermoso, F. Martinez-Boza, and C. Gallegos

Department of Chemical Engineering Technology Research Center of Chemical Products and Processes (Pro² TECS), University of Huelva, Faculty of Experimental Sciences, Campus del Carmen, 21071 Huelva, Spain

ABSTRACT

This work deals with the effect of viscosity modifier nature and concentration on the rheological properties of model oil-based drilling fluids (OBM) submitted to high pressure. The oil-based fluids were formulated by dispersing, with a high shear mixer, two selected organobentonites in a mineral oil, at room temperature. The viscous flow behaviour of the corresponding dispersions was characterised as a function of pressure, organoclay nature and organoclay concentration, using a controlled-stress rheometer equipped with both pressure cell and coaxial cylinder geometries. A factorial Sisko–Barus model, which takes into account both shear and pressure effects in the same equation, fitted the experimental pressure–viscosity data fairly well.

The influence of disperse phase concentration on the shear-thinning characteristics of these organoclay dispersions is related to the development of different microstructures, which depend on organoclay nature. In this sense, the resulting microstructure has been attributed to the cohesion energy between microgels domains. From the experimental results obtained, it can be concluded that the viscous flow behaviour of the OBM investigated is strongly affected by organoclay nature and concentration. The pressure–viscosity behaviour of these dispersions is mainly influenced by the piezoviscous properties of the oil and the properties of the continuous phase. The Sisko–Barus model proposed can be a useful tool, from an engineering point of view, for calculating pressure losses in the different sections of the bore, as well as being of significant help to solve other additional problems, such as hole cleaning, induced fracturing, and hole erosion during the drilling operation.

INTRODUCTION

Oil-based drilling fluids called oil based muds (OBM) are dispersions usually showing a complex rheology. Regarding the nature of their continuous phase, fluids used in drilling and completion wells can be classified into two main groups, water-based and oil-based. The main functions of these fluids are: i. To carry cuttings from the bottom of the hole, transport them up and remove rock bit at the surface. ii. To cool and clean the drill and the bit. iii. To maintain the stability of borehole. iv. To lubricate the gap between the drilling string and the wall of the hole. v. To prevent the inflow of fluids from surrounded rocks. vi. To form a thin and low-permeable filter cake. vii. To be non-damaging to the producing formation. viii. To be non-hazardous to the environment and personnel (Chilingarian and Vorabutr, 1983 and Menezes et al., 2010).

Most of global drilling operations use water-based muds (WBM), because of their lower environmental impact, whereas only 5–10% of the wells drilled use OBM (Caenn and Chillingar, 1996 and Meng et al., 2012). Nevertheless, OBM have interesting features to overcome certain undesirable characteristics of the water-based ones, such as better lubrication and higher boiling points (Khodja et al., 2010). Basically, OBM can be classified into three categories: (1) All-oil muds, consisting of a mixture of organoclay (OC) and synthetic or mineral oil, which are used for minimum pressure losses and low permeability reservoirs; (2) Oil muds, consisting of OC, emulsifiers, oil and water (2–10 m%), which are designed for well stabilisation at high temperature; (3) Invert oil muds, consisting of OC, emulsifiers, oil, additives and water (up to 40 m%), which are used for shale stability and improved penetration.

The relationship between flow behaviour and composition is an important issue to formulate suitable OBM. Regarding composition, clays are a key component for developing some specific properties of these dispersions, which will be submitted to the extreme pressure and temperature conditions of the wellbore. Concerning OBM, organophilic bentonites have been extensively used due to both good dispersing properties in the oil phase and filtration characteristics (Jordan et al., 1965). These OC result from the reaction of smectite-type and amine cationic groups, without addition of supplementary additives (Hauser, 1950 and Jordan, 1949).

Regarding flow properties, OBM are frequently submitted to extreme shear, temperature and pressure conditions in downhole operations. During the circulation of the drilling fluid around the wellbore, the shear rate may vary from zero to more than 1000

s^{-1}, whilst temperature can vary from values below 5 °C in water settings to above 200 °C at the bottom during the round trip. In addition, the pressure exerted by the mud column may be as much as 1400 bar at the deepest part (Darley and Gray, 1988). These severe conditions may change the bulk rheological behaviour of the dispersions because of pressure–temperature-dependent viscosity changes and particle–particle interactions modifications (Briscoe et al., 1994).

Several authors have examined the evolution of WBM viscosity with pressure and temperature, temperature being the most important factor (Santoyo et al., 2001 and Wang et al., 2010). Furthermore, other studies have been mainly focused on the use of new additives as rheology-modifiers to improve drilling operation. With this aim, recent studies have concentrated their efforts on how to overcome the hole instabilities, related to the aqueous media at extreme conditions, by using different additives, such as viscoelastic surfactants or synthetic polymers (England and Parris, 2010 and Wang et al., 2011).

For OBM, studies concerning the effect that pressure exerts on both their rheological behaviour and physical properties are very scarce. Combs and Whitmire (1968) studied the effect of temperature and pressure on the rheology of OBM formulated with OC, and found that the change in continuous phase viscosity was the main controlling factor. Politte (1985) concluded that the plastic viscosity could be normalised using the viscosity of the oil medium, whereas the yield stress is a weak function of pressure. Besides, Houwen and Geehan (1986) found a simple model to determine both yield-stress and high-shear-rate viscosity of invert muds as a function of pressure and temperature, using up to four parameters. In most cases, changes observed in physical properties and flow behaviour of OBM have been explained on the basis of the effect that both temperature and pressure exert on the viscosity of the continuous phase (Gandelman et al., 2007 and Herzhaft et al., 2001). Much less attention has been devoted to the effect that nature and concentration of viscosity modifiers exert on the rheological properties of oil dispersions submitted to high pressure (Ghalambor et al., 2008), probably due to the experimental constraints involving high pressure rheology measurements with fluids that exhibit non-Newtonian behaviour. Consequently, the overall objective of this work was to study the effect that viscosity modifier nature and concentration exert on the rheological properties of model OBM submitted to high pressure.

EXPERIMENTAL

Materials

Two commercially available OC, denoted as B34 and B128 and provided by Elementis (Belgium), were used in the present study. Their chemical formula and some physical characteristics are shown in Table 1.

Table 1: OC used in this study

Commercial name	Clay mineral	Abbreviated notation for intercalated ions[a]	Chemical formula	d_{001} (nm) [b]
Bentone 34	Bentonite	2M2HT	HT — N⁺ — HT with CH₃ above and CH₃ below	2.767
Bentone 128	Bentonite	2MBHT	CH₃ — N⁺ — CH₂—⟨benzyl⟩ with CH₃ above and HT below	3.344

[a]The abbreviations of quaternary ammonium ions corresponds to: M: methyl, B: benzyl, HT: hydrogenated tallow.

[b]Basal spacing determined by X-ray diffraction (XRD).

A mineral based lubricating oil, SR-10 (916 kg/m³ and 115 cSt, at 40 °C) supplied by Verkol (Spain), was used as base oil for the formulation of OBM.

Samples Preparation

Organobentonite dispersions were prepared by mixing OC (at concentration of 1, 3 and 5 m%.) in SR-10 oil base, at room temperature, using a high mixer Ultraturrax (Ika, Germany), at a rotational speed of 9000 rpm for five minutes. Prior to high shear processing, the OC were wetted with the oil, at room temperature, in a low shear mixer using a conventional four blade impeller.

X-Ray Diffraction (XRD)

XRD measurements were carried out on OC powders and their oily dispersions, at room temperature, using a Bruker S8 Advance (Germany) diffractometer equipped with a secondary monochromator, a Brentano Bragg geometry goniometer and a copper cathode as X-ray source. The samples were subjected to Cu K radiation with a wavelength of 0.15406 nm. The 2 angles varied from 1.5° to 20°, single scanning step of 0.017°, and measurement time of 6 s per step. The high intensity peaks in XRD curves show the d_{001}-spacing, which have been included in the Bragg equation to determine interlayer distance of each organically modified bentonite.

Optical Microscopy

Optical microscopy observations were carried out by using an Olympus BX52 (Japan) microscope, equipped with an Olympus C5050Z camera and an objective of 20× and 50× . An electric heating system LTS 350 (Linkam, UK) coupled with microscope stand was used to maintain the temperature constant. The OC dispersions were carefully poured into a sample holder and spread under the glass cover slip at room temperature. Before observations, all samples were heated up to 40 °C to compare both optical and rheological results.

Viscous Flow Measurements

Viscous flow measurements were performed using a controlled-stress rheometer, MARS II from Thermo-Scientific (Germany). Rheological data were obtained using a coaxial cylinder geometry (41 mm inner diameter, 1 mm gap, 60 mm length) at atmospheric pressure, and a coaxial cylinder-pressure cell D400/200 at high pressure. The cell D400/200 is a pressure vessel of 39 mm of inner diameter. Inside the cell, an inner cylinder of 38 mm diameter and 80 mm length was put in contact with a sapphire surface at the bottom of the vessel by a steel needle. This inner cylinder was equipped, at the top, with a secondary magnetic cylinder (36 mm diameter, 8 mm length), magnetically coupled to a tool outside the cell, which was connected to the motor-transducer of the rheometer. The pressure cell was connected to a hydraulic pressurisation system through a needle control valve.

A pressure transducer GMH 3110 (Gresingeg Electronic, Germany), able to measure differential pressures ranging 0 to 400 bar (0.1 bar resolution), was used.

Both atmospheric and high pressure rheological measurements were performed at 40 ± 0.1 °C using a circulating silicone bath.

Steady-state flow curves were obtained without sample pre-shear. The measurements were carried out by applying an increasing shear rate ramp, in the range comprised between 0.01 and 1000 s^{-1}. Two replicates of each rheological test were performed on fresh samples. The experimental error in viscosity was always inferior to ± 5%. Due to the fact that the viscous flow measurements might be influenced by a sedimentation process during the experimental flow time (Tropea et al., 2007), sedimentation rates were tested, at the above-mentioned temperature, to ensure that the flow measurements were not affected by phase separation.

RESULTS AND DISCUSSION

Viscous Flow Behaviour of OBM: Effect of OC Nature and Concentration

Viscous flow curves for the base oil and the drilling oil dispersions formulated with B34 and B128 OC, respectively, are shown in Fig. 1 and Fig. 2. As can be observed, OC dispersions (1–5 m%) display a shear-thinning behaviour with a tendency to reach a high-shear-rate-limiting viscosity. This shear-thinning behaviour is more apparent as OC concentration increases.

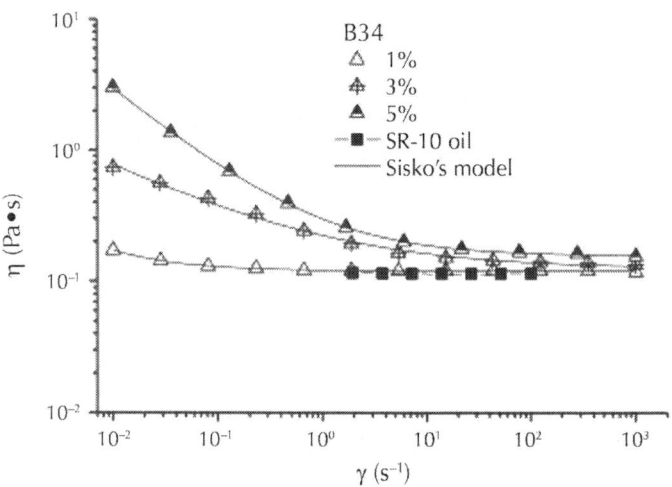

Figure 1: Viscous flow curves of OBM as a function of B34 OC concentration, at atmospheric pressure and 40 °C.

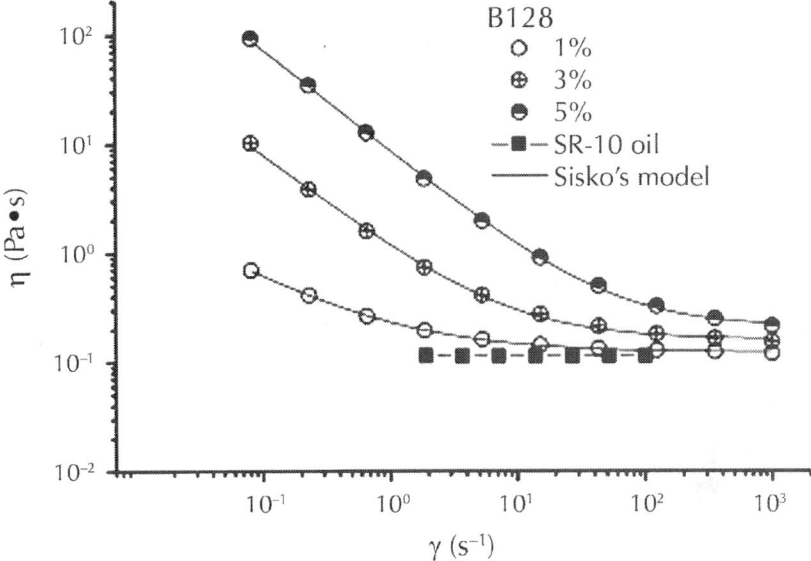

Figure 2: Viscous flow curves of OBM as a function of B128 OC concentration, at atmospheric pressure and 40 °C.

As can be seen in Fig. 1 and Fig. 2, Sisko›s model (Turian et al., 1998, Turian et al., 2002 and Weir and Bailey, 1996) fits the viscous flow behaviour of these OC (1–5%) dispersions fairly well:

$$\eta = \eta_{\infty 0} + k_0 \dot{\gamma}^{\eta_0 - 1}$$

(1)

where η is the apparent viscosity, $\eta_{\infty 0}$ is the high-shear-rate-limiting viscosity, k_0 the consistency index, and n_0 the flow index. Sisko's parameter values are shown in Table 2. The values of the average absolute relative deviation (%AARD), which is defined as:

$$\%AARD = \frac{100}{n} \sum_{i=1}^{n} \left| \frac{\eta_{i,\,exp} - \eta_{i,cal}}{\eta_{i,cal}} \right|$$

(2)

are also shown in Table 2.

Table 2: Some rheological parameters of the drilling fluids studied, as a function of OC concentration and pressure

| % wt. | $k = k_o + k_1\Delta P$ | | $n = n_o + n_1\Delta P$ | | η_∞ (Pa s) | β (bar^{-1}) | AARD (%) |
	k_o (Pa s^n)	k_1 (Pa s^n bar^{-1})	n_o	n_1 (bar^{-1})			
B128							
1	0.17	$-6.9E-5$	0.65	$1.2E-5$	0.100	0.0027	10.1
3	1.27	$-6.8E-4$	0.16	$3.4E-5$	0.153	0.0027	5.11
5	9.55	$-1.1E-2$	0.09	$5.6E-5$	0.195	0.0028	3.03
B34							
1	0.0081	$1.0E-4$	0.65	$-9.5E-4$	0.118	0.0027	1.77
3	0.097	$6.3E-4$	0.55	$-9.4E-4$	0.129	0.0027	11.6
5	0.14	$4.7E-4$	0.36	$-3.9E-4$	0.163	0.0027	6.01
Oil					0.114	0.0027	

The shear-thinning behaviour, observed for these dispersions, can be explained as a particular characteristic of elastic soft solids (King et al., 2007), especially in the case of concentrated dispersions. Thus, the rheological behaviour of these materials has been related to both dispersion state and volume fraction of the disperse phase, which could be modified by physical and/or chemical interactions between the OC molecules and oil medium. In this sense, Moraru (2001) investigated the influence that the type of OC exerts on the shear-thinning characteristics of non-aqueous media. Primarily, the rheological behaviour of these non-polar colloidal systems was affected by its morphology, related to different aggregation states of the clay from nano to macroscale. Taking into account that organobentonites are clays partially covered by alkylammonium molecules adsorbed at their surface, the structure and, consequently, the flow behaviour of these dispersions may be related to the interactions developed between the organophilic ions and the solvent, the organic chain density between platelets and the chemical nature of the medium (Le Pluart et al., 2004). These interactions, which normally increase with clay concentration, lead to an increase in viscosity, as has been pointed out elsewhere (Das Kanungo and McAtee, 1986).

In this sense, a remarkable increase in viscosity with clay concentration is shown in Fig. 1 and Fig. 2, for both OC, more important for B128 dispersions, indicating that an increase in shear rate could yield a progressive alignment of the platelets, leading to a pronounced shear-thinning behaviour (Massinga et al., 2010).

The different viscosity values observed in the low shear rate region for dispersions of similar concentration (higher values for B128 dispersions) suggest that the microstructure developed in both OC should be completely different. Previous results pointed out that the decrease in the flow index could be related to a higher degree of intercalation in clay platelets (Wagener and Reisinger, 2003). According to this, the flow behaviour observed for B128 dispersions, which present lower flow indexes (0.09–0.64) than B34 dispersions, seems to indicate that oil molecules easily penetrate into the interlayer space of the B128 OC, yielding a stronger structural network for this OC (Hato et al., 2011 and Zhang et al., 2003). In addition, it can be deduced from Fig. 1 and Fig. 2 that, for B34 dispersions, the high-shear-rate-limiting viscosity is achieved at lower shear rate, indicating the development of a weaker structure.

As has been previously reported, the rheology of a dispersion is a complex function of several factors, such as disperse phase volume fraction, particle shape and particle interactions (Mueller et al., 2010). At high shear rates, the viscosity values of the dispersions studied in this research mainly depend on the disperse phase concentration and not on the strength of the interactions between particles and continuous phase (Liang et al., 2011). The shear-thinning behaviour found, and its tendency to reach a high-shear-rate limiting viscosity close to the oil medium viscosity, reveal that dispersion microstructure is highly susceptible to shear, yielding a complete disruption at high shear rates (Ten Brinke et al., 2007).

These results clearly point out that both concentration and OC nature determine the rheological properties of the OBM studied, being the influence of both variables on dispersion flow behaviour a key criterion for the mud industry, in order to define downhole conditions, such as penetration rate, adequate viscosity to lift cuttings, hole cleaning, and prevention of excessively high strengths.

Viscous Flow Behaviour of OBM: Effect of Pressure

The viscous flow curves for B34 and B128 organobentonites dispersions as a function of pressure (1–390 bar), organobentonite nature and disperse phase concentration are shown in Fig. 3. In this sense,Fig. 3A and B gather the viscous flow curves for the lowest organobentonite concentration, whilst Fig. 3C and D collect the ones obtained for the highest organobentonite concentration.

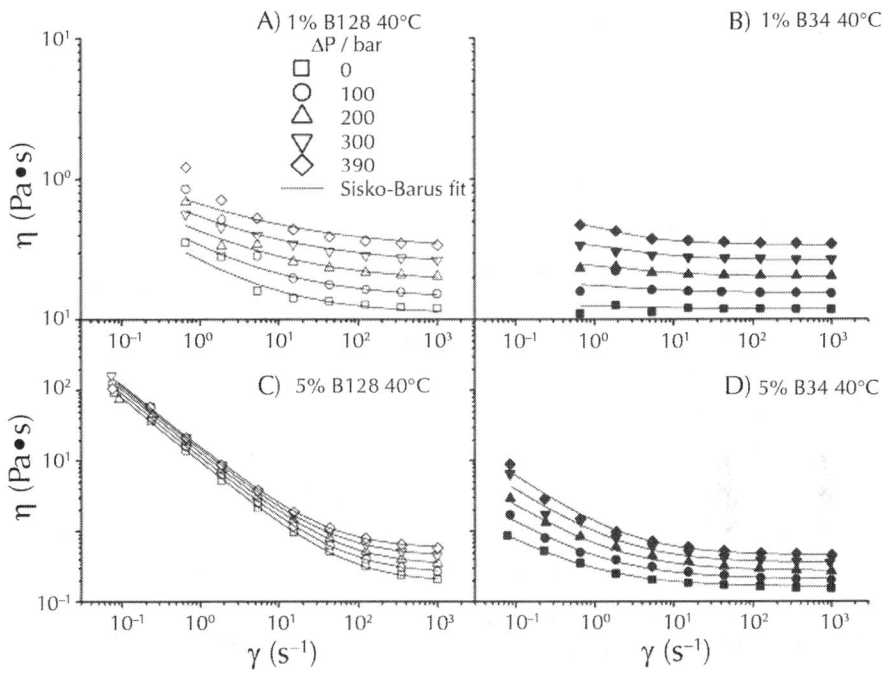

Figure 3: Experimental viscous flow curves, and Sisko–Barus' model fitting, for the different organobentonite dispersions studied, as a function of pressure, at 40 °C.

As expected, viscosity increases with pressure in the whole range of shear rate tested (Herzhaft et al., 2001). Likewise, the viscous flow behaviour of B34 dispersions is largely modified by an increase in OC concentration, from a quasi-Newtonian behaviour at the lowest concentration (1 m%, Fig. 3B), to an apparent shear-thinning response for the highest concentration (5 m%, Fig. 3D). On the contrary, dispersions made from B128 organobentonite show a clear shear-thinning behaviour in the whole range of disperse phase concentration studied.

In summary, the viscous flow behaviour of these OC dispersions significantly depends on pressure. This dependence can be modelled by using a modified Sisko model that includes the influence of pressure. In this case, Barus› model (Barus, 1893), which satisfactorily describes the pressure dependence of the oil viscosity at constant temperature in this range of pressure, can be used, in combination with Sisko›s model, to take into account both effects in the same equation. The expression, named Sisko–Barus› model, is given as:

$$\eta = \left[\eta_{\infty 0} + k(P)\dot{\gamma}^{n(P)-1}\right] \cdot \exp(\beta(P-P_0)) \tag{3}$$

being both consistency, $k(P)$, and flow indexes, $n(P)$, linear functions of pressure:

$$k(P)=k_0+k_1(P-P_0) \tag{4}$$

$$n(P)=n_0+n_1(P-P_0) \tag{5}$$

where $\eta_{\infty 0}$ is the high-shear-rate-limiting viscosity, at the reference pressure and 40 °C; β is the piezoviscous coefficient at 40 °C; P is the applied pressure; k_0 and n_0 are the consistency and flow indexes at the reference pressure ($P_0 = 1$ bar), respectively; and k_1, and n_1 are fitting parameters (bar^{-1}). Sisko–Barus' model fitting parameters have been gathered in Table 2.

As can be seen in Fig. 3A–D, the Sisko–Barus model describes OC dispersion viscosity evolution with both shear rate and pressure fairly well. It is interesting to note that the piezoviscous coefficients of the OBM analysed are quite similar to that of the oil base.

The effect of pressure on the rheology of dispersed system may be attributed to the physical changes that pressure exerts on both continuous and dispersed phases (Combs and Whitmire, 1968 and Hiller, 1963). In this case, OC addition does not seem to significantly affect the bulk piezoviscous properties of the continuous phase, despite the fact that oil phase compression implies a denser particle dispersion, being the effect of pressure on particles negligible as compared to the effect on the continuous oil medium, as has been explained elsewhere (Briscoe et al., 1994). Consequently, continuous phase volume reduction would be the main contribution to the piezoviscous coefficient in these dispersions.

Fig. 4 displays the evolution of the consistency index, for both organophilic bentonite dispersions, with pressure, in the range comprised between 0 and 390 bar, at 40 °C. It is worth remarking that parameter k_1 indicates a more or less important change in the consistency index with pressure.

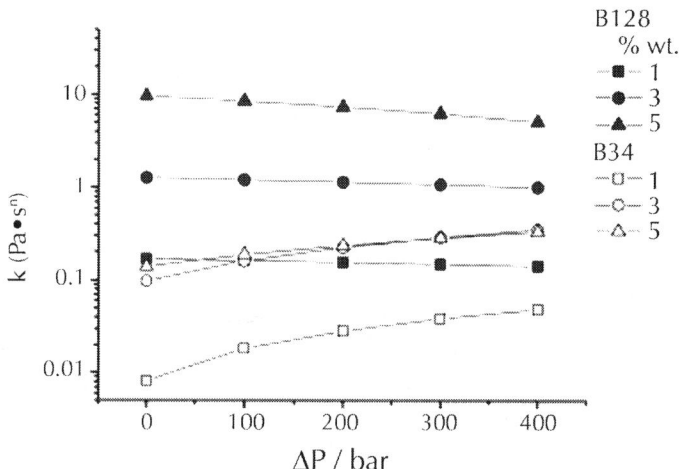

Figure 4: Evolution of the consistency index with pressure, as a function of OC nature and concentration (filled symbols: B128 OC dispersions; open symbols: B34 OC dispersions), at 40 °C.

OBM formulated with B34 OC exhibit a similar evolution of the pressure–viscosity curves for OC concentrations above 3 m%, with no significant differences in the consistency indexes, in the range of pressures studied. The positive values of k_1 would indicate increasing interparticle interactions, attributed to volume changes in the liquid phase under pressure, resulting in an effective increase in solid concentration.

On the other hand, OBM formulated with B128 always show a decrease of the consistency index with pressure ($k_1 < 0$, see Table 2). This fact suggests that B128 dispersions microstructure would consist of cohesive aggregates, whose strength considerably dampens the influence of oil compression on the bulk rheological response. Probably, a higher compatibility between B128 OC and oil promotes percolated structures, more robust against pressure changes than those developed for B34-based dispersions (Burgentzlé et al., 2004), where a more easily compressible non-packed structure, interconnected by weak interactions, would be developed.

The evolution of the flow index with pressure for the drilling fluid samples studied is shown in Fig. 5. As can be observed, the flow indexes of B34 OC dispersions, at 40 °C, linearly decrease as pressure increases (negative values of the flow index parameter n_1 with pressure). Besides, it can be also observed that the most concentrated dispersion (5 m%) shows the highest shear-thinning degree. Changes in the shear-thinning characteristics of dispersions with pressure have been also reported by Alderman et al. (1988), who found that the flow index slightly decreased with pressure, at 40 °C, for WBM, and by Houwen and Geehan (1986), which sustained that, for OBM, the viscous behaviour is largely determined by the physical properties of the continuous phase.

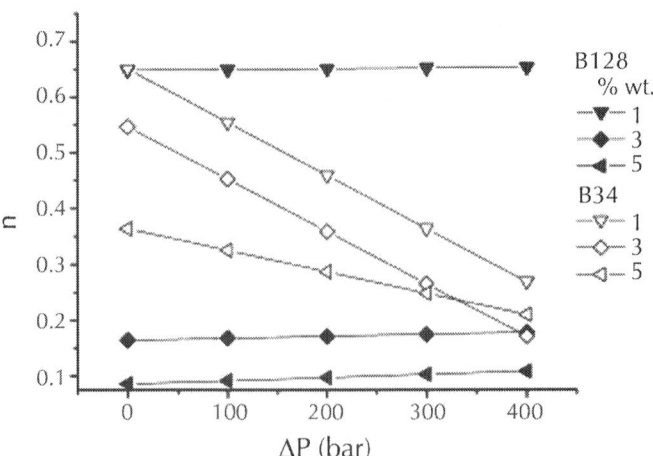

Figure 5: Evolution of the flow index with pressure, as a function of OC nature and concentration (filled symbols: B128 OC dispersions; open symbols: B34 OC dispersions), at 40 °C.

For B34-based OBM, the remarkable pressure dependence of the flow index suggests that a higher number of interparticle contacts could be induced by pressure packing. Thus, the dispersion with the largest effective volume fraction has the lowest flow indexes.

On the contrary, B128 dispersions behave completely different, as shown in Fig. 5. Thus, the flow indexes of these dispersions dramatically depend on OC concentration, whilst it is almost independent of pressure. In this case, the flow index is not affected by pressure, since the microstructure would be strong enough to resist changes in pressure. In this sense, the slight influence of pressure on shear-thinning characteristics for B128 dispersions is in agreement with previous results reported by Alderman et al. (1988) for water-based bentonite dispersions, showing that, for strong particle–particle interactions, changes in continuous phase rheology have a small relative influence on the bulk rheology of the dispersion (Briscoe et al., 1994).

The influence of pressure on viscosity, at both intermediate (10 s^{-1}) and large (1000 s^{-1}) shear rates, and at 40 °C, for selected OBM samples (1% and 5 m %), is illustrated in Fig. 6A–D. As can be observed inFig. 6C and D, the effect of concentration on fluid piezoviscous coefficient, at high shear rate, does not depend on the type of OC used in the formulation of the fluid, showing very similar piezoviscous coefficients to that of the base oil. On the contrary, in the low shear rate region (Fig. 6A and B), B128 OBM show larger increases in viscosity with OC concentration in relation to B34 OC containing dispersions, although, however, its piezoviscous coefficients decrease as OC concentration increases.

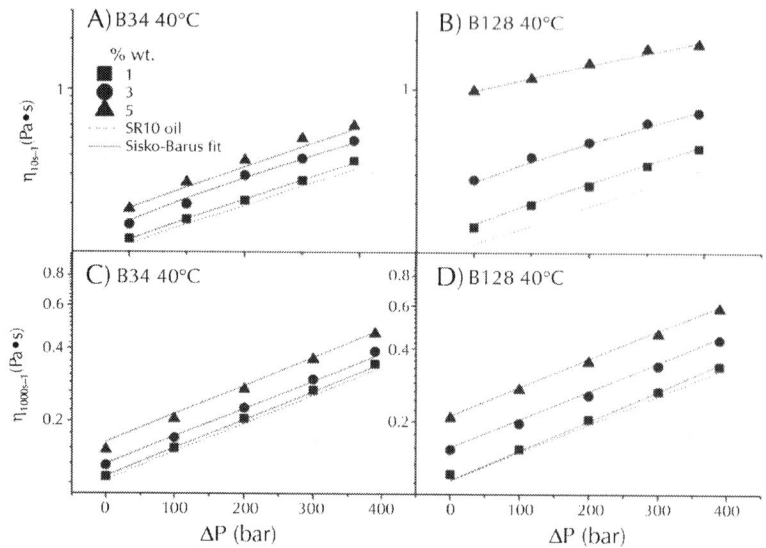

Figure 6: Influence of pressure on OBM apparent viscosities, at 40 °C, as a function of OC nature and concentration (symbols correspond to experimental values; solid lines correspond to Sisko–Barus› fitting). A and B: test carried out at 10 s⁻¹; C and D: tests carried out at 1000 s⁻¹.

On the other hand, the Sisko–Barus model proposed here describes fairly well the pressure dependence of viscosity, for both OC dispersions, at different shear rates. This model can be a useful tool, from the engineering point of view, to calculate pressure losses in the different sections of the bore, as well as being of significant help to solve different problems, i.e. hole cleaning, induced fracturing, and hole erosion during the drilling operation (Abdo and Haneef, 2012, Morita et al., 1996 and Whitfill et al., 2006).

OBM: Relationship between Dispersion Microstructure and Viscous Flow Behaviour

X-ray diffraction is a powerful technique in order to characterise the structure of clays, OC and composites from a nanoscale point of view (Pavlidou and Papaspyrides, 2008, Ras et al., 2007, Slade and Gates, 2004 and Starodoubtsev et al., 2006). XRD profiles for organobentonite dry powders (B34 and B128) and selected OBM dispersions (5 m% OC) are shown in Fig. 7.

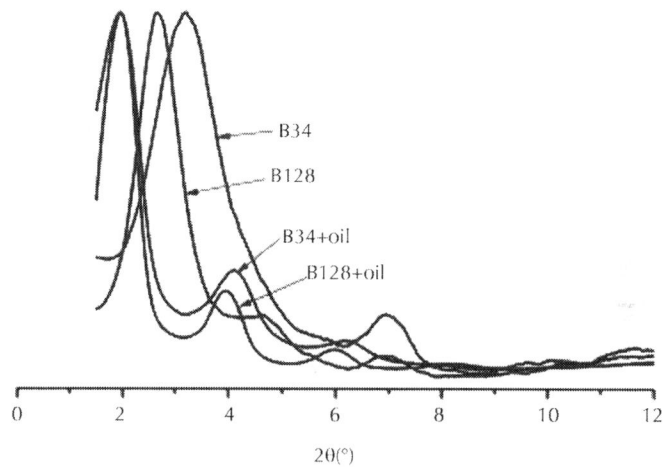

B34

B128

B34+oil

B128+oil

$2\theta(°)$

Figure 7: XRD patterns of OC dry powders (B34 and B128) and selected OBM (5 m% OC).

The differences observed in the diffraction patterns (d_{001}) of both OC (0.577 nm) can be attributed to the different arrangements of organic molecules between the layers. Furthermore, a comparison of these diffractograms with those obtained by Minase et al. (2008) reveals that interlayer structure of alkyl chains strongly depends on the amount of cations adsorbed. Thus, it has been estimated that more than 1.41 mmol of ammonium per gram of clay are present in these organobentonites. In this sense, the interlayer conformations developed by these organic groups could be interpreted as a paraffin-like bilayer structure allowing better fitting of the ammonium groups (Minase et al., 2008).

The presence of peaks at low angles $(d_{001} = 4.506$ nm) in both OC dispersions confirms that the oil yields an increase in interlayer spacing. This demonstrates the compatibility of the modified clays with the oil medium. The remarkable increase in interlayer spacing

(d_{001}) observed for both OC explains the increase in dispersion viscosity (see Fig. 1 and Fig. 2) as compared to base oil viscosity. Similar results have been reported by other authors (Hyun et al., 2001).

Fig. 7 also displays additional peaks. Thus, a second order peak (d_{002}) at 1.274 nm, more pronounced for B34 bentonite, and a third order peak, located at 1.430 and 1.479 nm, for B34 and B128 dispersions, respectively. Presumably, the third peak could be related to the formation of monolayer structures. Monolayer structures are possible if the area required for the flat-lying cations, with n carbons atoms in the chain, becomes the same as the area available to each univalent cation in a monolayer between the two silicate layers (Bonczek et al., 2002 and de Paiva et al., 2008). In spite of the fact that both B34 (n-metyl-alkyl ammonium) and B128 (benzyl ammonium) OC show the same maximum swelling peak (d_{001}), the second peak (d_{002}) presents a slight shifting. This result, in addition of the existence of a third peak corresponding to d_{003}, could indicate that the oil creates different disordered structures in both bentonite dispersions, probably due to the partial solvation of hydrocarbons. These materials would form paraffin-like bilayer structures and disordered conformations, such as pseudotrimolecular layers, as has been pointed out elsewhere (He et al., 2004). Additionally, the lower values of second and third order diffraction picks of B128 dispersion could be related to a higher affinity between the aromatic radical group of this OC and the naphthenic nature of the oil (Burgentzlé et al., 2004 and Paul and Robeson, 2008) and, consequently, yielding a more solid-like configuration of the interlayer structures. Although XRD provides a qualitative structural information for elucidate interlayer spacing, the similar intercalation of the oil molecules observed in both types of dispersions does not explain the different rheological properties observed in these dispersions (Burgentzlé et al., 2004), suggesting that d-spacing swelling parameter is not the more significant variable to predict the viscous flow behaviour.

The swelling of organophilic clays can favour the formation of gels, having a remarkable elasticity, due to several factors, such as the interactions developed between the organophilic ions and the solvent, alkyl chain length and nature of organic groups, and chemical nature of the continuous medium (Gherardi et al., 1996, Jordan, 1949 and Moraru, 2001), among others.

In order to gain some insight into the microstructure–rheology relationship of these OBM, optical micrographs for both B34 (A, B, C) and B128 dispersions (D, F, E) are depicted in Fig. 8. It can be observed that the OBM formulated with B34 OC display a large number aggregated particles dispersed in the continuous oil background. Particle volume fraction and size of these aggregates increase with OC concentration (Wagener and Reisinger, 2003). In contrast, no agglomerated particles can be observed at any concentration for B128 dispersions.

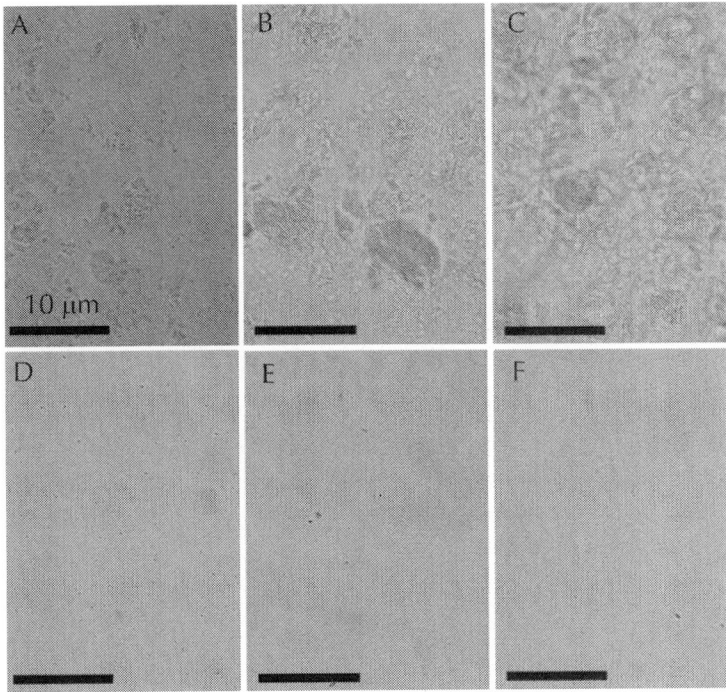

Figure 8: Optical micrographs, at 40 °C, for the different OC dispersions studied. A, B, C: B128 OC dispersions (1, 3 and 5 m%); D, E, F: B34 OC dispersions (1, 3 and 5 m%).

The results from XRD and optical microscopy suggest that the microstructure of these dispersions presumably consists of particles and aggregates (King et al., 2007). At the microscopic scale, particles lead to different microstructures depending on the organic ions/solvent interactions. Hence, the dimetyl alkyl covered clay, easily swell, leading to aggregates consisting of multipacked anisotropic shaped particles, probably due to short-range attractive force which prevents disaggregation (Moraru, 2001), not having enough affinity to connect all the aggregates into a percolated assembled structure. Consequently, an increase in OC concentration favours the formation of larger aggregates, as can be observed in Fig. 8, without showing dramatically higher viscosities.

On the contrary, the benzyl ammonium ion in B128 OC yields a significant gel-like behaviour, as a consequence of strong interactions among clay surface and hydrocarbons molecules, due to a better solubility of the benzyl ammonium ion in the naphthenic oil. This OC probably develops a microstructure of tiny aggregates through thin liquid interlayers with consistent linkage points between the tactoids of the physical network (Moraru, 2001), a fact that would explain the higher viscosity values as well as the stronger shear-thinning behaviour of B128 dispersions compared to B34 ones (Krishnamoorti et al., 1996), especially for high OC concentration.

Of course, the above-mentioned microstructure influences dispersion rheology as a function of pressure. For B34 dispersions, the effect of pressure on the rheological parameters (k, n) of the Sisko–Barus model depends on OC concentration, resulting from the increase in the effective disperse phase volume fraction. In the case of B128 dispersions, the pressure dependence of the viscous behaviour is essentially independent of concentration, due to the significant gel connectivity. However, despite of the different microstructures found for both types of OBM, the effect of compression on viscosity is essentially dominated by the continuous oily phase, showing quite similar piezoviscous values for all the formulations tested.

CONCLUSIONS

From the experimental results obtained, it can be concluded that the viscous flow behaviour of the OBM investigated is strongly influenced by OC nature and concentration. The pressure–viscosity behaviour of these dispersions is mainly influenced by the piezo-viscous properties of the oil and the properties of the continuous phase. The Sisko–Barus model proposed in the present work predicts the evolution of the non-Newtonian viscosity of oil drillings fluids tested with pressure and shear rate fairly well. This model can be a useful tool, from the engineering point of view, to calculate pressure losses in the different sections of the bore, as well as being of significant help to solve other additional problems, such as hole cleaning, induced fracturing, and hole erosion during the drilling operation.

In spite of the fact that both OC display many similarities at the nanoscale level, the microstructures resulting from interactions among organic ions and solvent are quite diverse, yielding significant differences in the bulk viscous flow behaviour of their respective dispersions.

ACKNOWLEDGMENTS

This work has been sponsored by FEDER-Excellence Projects Programme (Research project P08-TEP-3895, Junta de Andalucía, Spain). The authors gratefully acknowledge the financial support.

REFERENCES

1.   Abdo, J., Haneef, M.D., 2012. Nano-enhanced drilling fluids: pioneering approach to overcome uncompromising drilling problems. J. Energy Resour. Technol. 134, 014501. http://dx.doi.org/10.1115/1.4005244.

2.   Alderman, N.J., Gavignet, A., Guillot, D., Maitland, G.C.,

1988. High-temperature, highpressure rheology of water based muds. Society of Petroleum Engineers of AIME, (Paper), 18035, pp. 187–195.

3.   Barus, C., 1893. Isothermals, isopiestics and isometrics relative to viscosity. Am. J. Sci. 45, 87–96.

4.   Bonczek, J.L., Harris, W.G., Nkedi-Kizza, P., 2002. Monolayer to bilayer transitional arrangements of hexadecyltrimethylammonium cations on Na-Montmorillonite. Clay Clay Miner. 50, 11–17.

5.   Briscoe, B.J., Luckham, P.F., Ren, S.R., 1994. The properties of drilling fluids at high pressure and high temperatures. Philos. Trans. R. Soc. Lond. A 348, 179–207.

6.   Burgentzlé, D., Duchet, J., Gérard, J.F., Jupin, A., Fillon, B., 2004. Solvent-based nanocomposite coatings I. Dispesions of oganophilic montmorillonita in organic solvents. J. Colloid Interface Sci. 278, 26–39.

7.   Caenn, R., Chillingar, G.V., 1996. Drillings fluids: state of art. J. Pet. Sci. Eng. 14, 221–230. Chilingarian, G.V., Vorabutr, P., 1983. Drilling and Drilling Fluids, Second ed. Elsevier, Amsterdam, Netherlands.

8.   Combs, G.D., Whitmire, L.D., 1968. Capillary viscometer simulates bottom-hole conditions. Oil Gas J. 108–113 (30 September).

9.   Darley, H.C.H., Gray, G.R., 1988. Composition and properties of drilling and completion fluids, Sixth ed. Gulf Publ. Co., Houston, USA.

10.   Das Kanungo, J.L., McAtee Jr., J.L., 1986. Effects of polymers and clay concentrations on the viscosities of organo-smectite dispersions under high pressure. Appl. Clay Sci. 1, 285–293.

11.   de Paiva, L.B., Morales, A.R., Velenzuela Díaz, F.R., 2008. Organoclays: properties, preparation and applications. Appl. Clay Sci. 42, 8–24.

12.   England, K.W., Parris, M.D., 2010. Viscosity influences of high pressure on borate crosslinked gels. SPE Deepwater Drilling and Completions Conference, Galveston, USA.

13. Gandelman, R.A., Leal, R.A.F., Gonçalves, Aragão, A.F., Lomba, R.F., Martins, A.L., 2007. Study on gelation and freezing phenomena of synthetic drilling fluids in ultradeepwater environments. SPE/IADC Drilling Conference and Exhibition, Amsterdam, Netherlands.

14. Ghalambor, A., Ashrafizadeh, S.N., Nasiri, M., 2008. Effect of basic parameters on the viscosity of synthetic-based drilling fluids. SPE International Symposium and Exhibition on Formation Damage Control, Lafayette, USA.

15. Gherardi, B., Tahani, A., Levitz, P., Bergaya, F., 1996. Sol/gel phase diagrams of industrial organo-bentones in organic media. Appl. Clay Sci. 11, 163–170.

16. Hato, M.J., Zhang, K., Ray, S.S., Choi, H.J., 2011. Rheology of organoclay suspension. Colloid Polym. Sci. 289, 1119–1125.

17. Hauser, E.A., 1950. Modified gel-forming clay and process of producing same. U.S. Patent No. 2,531,427.

18. He, H., Frost, R.L., Deng, F., Zhu, J., Wen, X., Yuan, P., 2004. Conformation of surfactant molecules in the interlayer of montmorillonite studied by 13C MAS NMR. Clay Clay Miner. 52, 350–356.

19. Herzhaft, B., Peysson, Y., Isambourg, P., Delepoulle, A., Toure, A., 2001. Rheological properties of drilling muds in deep offshore conditions. SPE/IADC Drilling Conference, Amsterdam, Netherlands.

20. Hiller, K.H., 1963. Rheological measurements of clay suspensions at high temperatures and pressures. J. Petrol. Technol. 15, 779–788.

21. Houwen, O.H., Geehan, T., 1986. Rheology of Oil-Base Muds. SPE Annual Technical Conference and Exhibition, New Orleans, USA.

22. Hyun, Y.H., Lim, S.T., Choi, H.J., Jhon, M.S., 2001. Rheology of poly(ethylene oxide)/ organoclay nanocomposites. Macromolecules 34, 8084–8093.

23. Jordan, J.W., 1949. Organophilic bentonites. I. Swelling in organic liquids. J. Phys. Chem. 53, 294–306.

24. Jordan, J.W., Nevins, M.J., Stearns, R.C., Cowan, J.C., Beasley, A.E., 1965. Well- working fluids. U.S. Patent No. 3,168,475.

25. Khodja, M., Canselier, J.P., Bergaya, F., Fourar, K., Khodja, M., Cohaut, N., Benmounah, A., 2010. Shale problems and water-based drilling fluid optimisation in the Hassi Messaoud Algerian oil field. Appl. Clay Sci. 49, 383–393.

26. King Jr., H.E., Milner, S.T., Lin, M.Y., Singh, J.P., Mason, T.G., 2007. Structure and rheology of organoclay suspensions. Phys. Rev. E. 75, 021403-1–021403-20.

27. Krishnamoorti, R., Vaia, R.A., Giannelis, E.P., 1996. Structure and dynamics of polymerlayered silicate nanocomposites. Chem. Mater. 8, 1728–1734.

28. Le Pluart, L., Duchet, J., Sautereau, H., Halley, P., Gerard, J.F., 2004. Rheological properties of organoclay suspensions in epoxy network precursors. Appl. Clay Sci. 25, 207–219.

29. Liang, R., Han, L., Doraiswamy, D., Gupta, R.K., 2011. The rheology of aramid platelet suspensions. Polym. Eng. Sci. 51, 1933–1941.

30. Massinga, P.H., Focke, W.W., de Vaal, P.L., Atanasova, M., 2010. Alkyl ammonium intercalation of Mozambican bentonite. Appl. Clay Sci. 49, 142–148.

31. Menezes, R.R., Marques, L.N., Campos, L.A., Ferreira, H.S., Santana, L.N.L., Neves, G.A., 2010.

32. Use of statistical design to study the influence of CMC on the rheological properties of bentonite dispersions for water-based drilling fluids. Appl. Clay Sci. 49, 13–20.

33. Meng, X., Zhang, Y., Zhou, F., Chu, P.K., 2012. Effects of carbon ash on rheological properties of water drilling fluids. J. Pet. Sci. Eng. 100, 1–8.

34. Minase, M., Kondo, M., Onikata, M., Kawamura, K., 2008. The viscosity of organic liquid suspensions of trimethyldococylammonium-montmorillonite complexes. Clay Clay Miner. 56, 49–65.

35. Moraru, V.N., 2001. Structure formation of alkylammonium

montmorillonites in organic media. Appl. Clay Sci. 19, 11–26.

36.   Morita, N., Black, A.D., Fuh, G.-F., 1996. Borehole breakdown pressure with drilling fluidsl. Empirical results. Int. J. Rock Mech. Min. Sci. Geomech. Abstr. 33, 39–51.

37.   Mueller, S., Llewellin, E.W., Mader, H.M., 2010. The rheology of suspensions of solid particles. Proc. R. Soc. A 466, 1201–1228.

38.   Paul, D.R., Robeson, L.M., 2008. Polymer nanotechnology: Nanocomposites. Polymer 49, 3187–3204.

39.   Pavlidou, S., Papaspyrides, C.D., 2008. A review on polymer-layered silicate nanocomposites. Prog. Polym. Sci. 33, 1119–1198.

40.   Politte, M.D., 1985. Invert oil mud rheology as a function of temperature and pressure. SPE/IADC Drilling Conference, New Orleans, USA.

41.   Ras, R.H.A., Umemura, Y., Johnston, C.T., Yamagishi, A., Schoonheydt, R.A., 2007. Ultrathin hybrid films of clay minerals. Phys. Chem. Chem. Phys. 9, 918–932.

42.   Santoyo, E., Santoyo-Gutiérrez, S., García, A., Espinosa, G., Moya, S.L., 2001. Rheological property measurement of drilling fluids used in geothermal wells. Appl. Therm. Eng. 21, 283–302.

43.   Slade, P.G., Gates, W.P., 2004. The swelling of HDTMA smectitec as influenced by their preparation and layer charges. Appl. Clay Sci. 25, 93–101.

44.   Starodoubtsev, S.G., Lavrentyeva, E.K., Khokhlov, A.R., Allegra, G., Famulari, A., Meille, S.V., 2006. Mechanism of smectic arrangement of montmorillonite and bentonite clay platelets incorporated in gels of poly(acrylamide) induced by the interaction with cationic surfactants. Langmuir 22, 369–374.

45.   Ten Brinke, A.J.W., Bailey, L., Lekkerkerker, H.N.W., Maitland, G.C., 2007. Rheology modification in mixed shape colloidal

dispersions. Part I: Pure components. Soft Matter 3, 1145–1162.

46. Tropea, C., Yarin, A.L., Foss, J.F., 2007. Springer handbook of experimental fluid mechanics, First ed. Springer Verlag, Berlin, Germany.

47. Turian, R.M., Ma, T.-W., Hsu, F.-L.G., Sung, D.-J., 1998. Flow of concentrated nonNewtonian slurries: 1. Friction losses in laminar, turbulent and transition flow through straight pipe. Int. J. Multiphase Flow 24, 225–242.

48. Turian, R.M., Attal, J.F., Sung, D.-J., Wedgewood, L.E., 2002. Properties and rheology of coalwater mixtures using different coals. Fuel 81, 2019–2033.

49. Wagener, R., Reisinger, T.J.G., 2003. A rheological method to compare the degree of exfoliation of nanocomposites. Polymer 44, 7513–7518.

50. Wang, F., Wang, R., Liu, J., Wang, L., Li, J., Che, L., Su, H., 2010. Rheology of high-density water-based drilling fluid at high temperature and high pressure. Shiyou Xuebao/ Acta Petrol. Sin. 31, 306–310.

51. Wang, J., Zheng, J., Musa, O.M., Farrar, D., Cockcroft, B., Robinson, A., Gibbison, R., 2011. Salt-tolerant, thermally-stable rheology modifier for oilfield drilling applications. SPE International Symposium on Oilfield Chemistry, The Woodlands, USA.

52. Weir, I.S., Bailey, W.J., 1996. Statistical study of rheological models for drilling fluids. SPE J. 1, 473–485.

53. Whitfill, D.L., Jamison, D.E., Wang, M., Thaemlitz, C., 2006. New design models and materials provide engineered solutions to lost circulation. SPE Russian Oil and Gas Technical Conference and Exhibition, Moscow, Russia.

54. Zhang, L.-M., Jahns, C., Hsiao, B.S., Chu, B., 2003. Synchotron SAXS/WAXD and rheological studies of clay suspensions in silicone fluid. J. Colloid Interface Sci. 266, 339–345.

Citations

CHAPTER 1

Iheoma M. Adekunle, Augustine O. O. Igbuku, Oke Oguns and Philip D. Shekwolo (2013). Emerging Trend in Natural Resource Utilization for Bioremediation of Oil — Based Drilling Wastes in Nigeria, Biodegradation - Engineering and Technology, Dr. Rolando Chamy (Ed.), ISBN: 978-953-51-1153-5, InTech, DOI: 10.5772/56526.

CHAPTER 2

Chuanliang Yan, Jingen Deng, and Baohua Yu, "Wellbore Stability in Oil and Gas Drilling with Chemical-Mechanical Coupling," The Scientific World Journal, vol. 2013, Article ID 720271, 9 pages, 2013. doi:10.1155/2013/720271.

CHAPTER 3

Liv A. Carlsen, Gerhard Nygaard, Michael Nikolaou, Evaluation of control methods for drilling operations with unexpected gas influx, Journal of Process Control, Volume 23, Issue 3, March 2013, Pages 306-316, ISSN 0959-1524, http://dx.doi.org/10.1016/j.jprocont.2012.12.003.

CHAPTER 4

K.A. Fattah, S.M. El-Katatney, A.A. Dahab, Potential implementation of underbalanced drilling technique in Egyptian oil fields, Journal of King Saud University - Engineering Sciences, Volume 23, Issue 1, January 2011, Pages 49-66, ISSN 1018-3639, http://dx.doi.org/10.1016/j.jksues.2010.02.001.

CHAPTER 5

Jana D. Abou Ziki, Rolf Wüthrich, Forces exerted on the tool-electrode during constant-feed glass micro-drilling by spark assisted chemical engraving, International Journal of Machine Tools and Manufacture, Volume 73, October 2013, Pages 47-54, ISSN 0890-6955, http://dx.doi.org/10.1016/j.ijmachtools.2013.06.008.

CHAPTER 6

Ivica Kisic, Sanja Mesic, Ferdo Basic, Vladislav Brkic, Milan Mesic, Goran Durn, Zeljka Zgorelec, Lidija Bertovic, The effect of drilling fluids and crude oil on some chemical characteristics of soil and crops, Geoderma, Volume 149, Issues 3–4, 15 March 2009, Pages 209-216, ISSN 0016-7061, http://dx.doi.org/10.1016/j.geoderma.2008.11.041.

CHAPTER 7

J. Hermoso, F. Martinez-Boza, C. Gallegos, Influence of viscosity modifier nature and concentration on the viscous flow behaviour of oil-based drilling fluids at high pressure, Applied Clay Science, Volume 87, January 2014, Pages 14-21, ISSN 0169-1317, http://dx.doi.org/10.1016/j.clay.2013.10.011.

Index